RATIONAL CHANGES IN SCIENCE

BOSTON STUDIES IN THE PHILOSOPHY OF SCIENCE

EDITED BY ROBERT S. COHEN AND MARX W. WARTOFSKY

VOLUME 98

RATIONAL CHANGES IN SCIENCE

Essays on Scientific Reasoning

Edited by

JOSEPH C. PITT

*Department of Philosophy, Virginia Polytechnic Institute
and State University*

and

MARCELLO PERA

*Dipartimento di Filosofia,
Università degli Studi di Pisa*

D. REIDEL PUBLISHING COMPANY

A MEMBER OF THE KLUWER ACADEMIC PUBLISHERS GROUP

DORDRECHT / BOSTON / LANCASTER / TOKYO

Library of Congress Cataloging in Publication Data

Modi del progresso. English.
    Rational changes in science.

    (Boston studies in the philosophy of science; v. 98)
    Translation of: I Modi del progresso.
    Includes index.
    1. Science – Philosophy. 2. Rationalism. I. Pitt, Joseph C.
II. Pera, Marcello, 1943–   . III. Title. IV. Series.
Q174.B67 vol. 98 [Q175]   501 s [501]   87–12808
ISBN-13:978-94-010-8181-8   e-ISBN-13:978-94-009-3779-6
DOI: 10.1007/978-94-009-3779-6

Published by D. Reidel Publishing Company,
P.O. Box 17, 3300 AA Dordrecht, Holland.

Sold and distributed in the U.S.A. and Canada
by Kluwer Academic Publishers
101 Philip Drive, Norwell, MA 02061, U.S.A.

In all other countries, sold and distributed
by Kluwer Academic Publishers Group,
P.O. Box 322, 3300 AH Dordrecht, Holland.

First published in Italian as *I modi del progresso*
in 1985 by il Saggiatore, Milan.

TABLE OF CONTENTS

THE PROBLEMS OF SCIENTIFIC RATIONALITY

Fashion is a fickle mistress. Only yesterday scientific rationality enjoyed considerable attention, consideration, and even reverence among philosophers; "but today's fashion leads us to despise it, and the matron, rejected and abandoned as Hecuba, complains; *modo maxima rerum, tot generis natisque potens — nunc trahor exul, inops*", to cite Kant for our purpose, who cited Ovid for his.

Like every fashion, ours also has its paradoxical aspects, as John Watkins correctly reminds in an essay in this volume. Enthusiasm for science was high among philosophers when significant scientific results were mostly a promise, it declined when that promise became an undeniable reality. Nevertheless, as with the decline of any fashion, even the revolt against scientific rationality has some reasonable grounds. If the taste of the philosophical community has changed so much, it is not due to an incident or a whim.

This volume is not about the history of and reasons for this change. Instead, it provides a view of the new emerging image of scientific rationality in both its philosophical and historical aspects. In particular, the aim of the contributions gathered here is to focus on the concept around which the discussions about rationality have mostly taken place: scientific change.

In debates over this subject it seems as if we have passed from one myth to another. In the beginning there was the myth of the data (MD) — as it is called here by Raimo Tuomela. This myth was based on two dogmas, one the counterpart of the other: — the empiricists' dogma of the immaculate perception — the rationalists' dogma of the immortality of the rational soul.

Change of scientific theories was not a big problem in MD because either the empirical data were taken as eternal or the conceptual data of the mind were considered as not affected by history. According to MD, these two elements represent a bridge between competing theories and a measure for common evaluation. Even Pierre Duhem, who had already shaken the traditional view of scientific rationality, maintained "according to a continuous tradition, every theory of physics passes to the next that section of natural classification that it was able to build".

J. C. Pitt and M. Pera (eds.), Rational Changes in Science, vii —xiii
© *1987 by D. Reidel Publishing Company.*

With the decreasing popularity of the two concrete dogmas, MD was transformed into the "myth of the framework", as it is called by Popper. No longer did we find continuity, but radical breaks instead. No longer were there firm bridges between theories, incommensurability between different conceptual frameworks was the agreed upon state of affairs. Furthermore, what previously had not been a problem became the most central and burning issue: how to introduce rational criteria into theory change.

In analyzing the non-rational consequences of the myth of the framework, appeal was made to "the scientific method". But in such an appeal there was the risk of inventing another myth, the myth of the method. In the specific case that matters to us, the appeal to method immediately raises two problems or two classes of problems. They are:

(1) does a rational method of theoretical change exist?

(2) do rational theoretical changes exist in the history of science?

The essays of philosophical and logical analysis that constitute the first part of this volume deal mostly with problem 1; the essays of historical investigation combined with philosophical analysis of the second part deal instead with problem 2.

But there is also a third problem:

(3) on what grounds is a methodology of theoretical change justified as rational?

While 1 and 2 run in parallel, as Burian points out, 3 crosses almost all the essays of this volume. We do not intend to discuss the merits of their proposed problems and solutions. Instead, in order to introduce the reader to the scope of the issue and the range of approaches to it, we prefer to offer a review. Even though the solutions presented are not univocal (which is not possible, nor was it our intention), they largely converge on some important points.

Let us start with problems 1 and 2. One of the most positive characteristics of the historical attitude of current philosophy of science is the analysis of science not only as a product but also as a process. As a consequence, research has expanded into areas previously considered taboo for philosophers. The so-called context of discovery, concerned with the generation and introduction of new ideas, has been the first to benefit. Pera's contribution is dedicated to this subject.

Beginning in the second half of the 19th century, the Baconian ideal (shared also by Descartes, Newton and Mill) of a *Novum Organon* (a logic of discovery) was replaced by the ideals advocated by Whewell and the hypothetico-deductivists, of a *Novum Organon Renovatum*. Those efforts to revive a logic of discovery were challenged by the claim that even if rules for discovery really existed, they wouldn't have epistemic strength *per se*. Pera attempts to meet this challenge. He argues that on the basis not only of logical and philosophical motives, but also through the reconstruction of an historical case (Galvani's discovery of animal electricity) we can show there is a way from facts to hypotheses, that the arguments associated with discovery are the same as those on which the first acceptance of an hypothesis is based. By including theoretical factors and paying attention to the interplay of general and contextual heuristic principles, the *Novum Organon Reformatum* project promises to explain not only the fallibility but also the success of research and the dialectics of continuity and innovation in theoretic change.

But how does this change take place? The starting point of William Shea's essay is an idea of Whewell's. Shea develops in detail the study of the historical case to which Whewell referred, the theory of the vortex. According to Shea, Whewell's idea was that "when a prevalent theory is found to be untenable, and consequently, is succeeded by a different, or even an opposite one, the change is not made suddenly, or completed at once, at least in the minds of the most tenacious adherents of the earlier doctrine; but is effected by a transformation, or series of transformations, of the earlier hypothesis, by means of which it is gradually brought nearer and nearer to the second;" Whewell also believed in the continuity of scientific progress. In his *History of the Inductive Science* he wrote that, in historical change, "the previous truths are not rejected but absorbed, not contradicted but expanded" and that "the history of every science that can look like a sequence of revolutions is, in reality, a set of developments." Today, given the work of Kuhn and Feyerabend, these ideas are no longer accepted. Nevertheless, there are cases where they are historically adequate. As Shea shows, the theoretical difficulties of the Cartesian theory of vortex which remained even after Huygens' efforts to mathematize it did not produce one giant step toward Newton's theory of action at a distance, but instead resulted in a series of compromises and intermediate adjustments. Shea argues this is not a unique episode. The *Almagestum Novum*, written in 1651 by G. B. Riccoli, represents another typical example of a compromise among alternative theories (geocentric and heliocentric theories). In cases like these, the substitution of one theory for another seems to obey a logic of slow assimilation rather than a logic of sudden conversion.

But does such a logic of assimilation exist? For those intermediate periods which precede the acquisition of stringent proofs, Shea claims the evaluations of preferences are more like "expressions of taste". Rachel Laudan provides us with an analysis of one of these periods. On her view, a theory goes through at least two phases after it has been proposed before being accepted by a community. These two phases, to which the liberalization of the epistemic contexts now confers full philosophical dignity, are: "the context of taking into account", which concerns the scientist's willingness to admit a new theory among the theories worthy of consideration, and "the context of pursuit", in which the scientist develops a theory through both the articulation of the theoretical basis and the collection of new supporting data. When the contexts were only two — discovery and justification — cases like those studied by Laudan were almost incomprehensible. In continental drift we find ourselves faced with a theory that had to wait half a century before its merits were recognized. Was this scientific behavior rational? Laudan answers affirmatively. According to her, it conforms to a set of general rational prerequisites for taking into consideration and for the pursuit of scientific theories.

However, even if reasonable, as Laudan herself recognizes, her conditions are not the only possible ones. New problems then crop up; for instance, couldn't cases of theoretical changes be contextualized to changes in the very criteria of rationality? Galileo's revolution, examined here by J. Pitt, is one such. Pitt shows that the rationality of the solution to a scientific problem depends not only on the appropriateness of the problem itself (the problem of the tides in this specific case) but also on prior agreement as to acceptable strategies or modes of proof and upon certain other assumptions regarding the criteria for explanations. Galileo's idea of explanation differed from that of his Aristotelian opponents. For Galileo an explanation required a demonstration, the exhibition of "necessary connections," plus an interpretation in physical terms according to a scheme that reminds us of the neopositivist idea of language as an interpreted calculus. The proof of a theorem of physics or applied geometry is the prototype of this account of explanation.

On the other hand, the use of the geometric method in providing explanations, which separates Galileo from the Aristotelians, and whose justification is, according to Pitt, the most important aim of the First Day of the *Dialogue*. is based on the conviction shared by both parties about the advantages of geometry: its clarity and rigor.

It would seem that, even in cases of deep changes, the appeal to common elements or to the same intellectual framework or to pieces of the same framework is a condition of rational evaluation and preferences. But are these rational preferences still possible when the frameworks are different? Can individuals belonging to different traditions understand each other? To deny the possibility of such communication is to subscribe to what Popper here calls "the myth of the framework," according to which "a rational and fruitful discussion is impossible unless the participants share a common framework of basic assumptions or, at least, unless they have agreed on such a framework for the purpose of the discussion". This myth, shown by Popper to be long-standing, is today supported, for example, by Kuhn's thesis of paradigm incommensurability and Quine's thesis of ontological relativity. Against these forms of relativism Popper aims to show that ontological relativity makes evaluation difficult but not impossible. Thanks to the confrontation among languages, traditions, and different cultures, "the speakers can free themselves from prejudices they are not aware of, from the granted and unaware acceptance of theories like thorns incorporated in the logical structure of their language".

However, in order for comparisons among frameworks to be possible and productive, there is at least one invariable feature that even Popper has to recognize: supporters of different frameworks must share what he calls "the correct method of criticism", namely a method which does not aim to justify or establish theses, but to criticize them. The problem here is how to justify this alleged methodological invariance. Is it merely a fact of the history of science or is it the consequence of other facts, for example of invariance in cognitive values (that is, a norm), or do invariances simply not exist?

We are now at the point where we can turn to problem 3, "the source or basis of the normative force of the methodological rules and recommendations" as Thomas Nickles defines it.

Popper has always supported a non-naturalistic view of method. According to Nickles, however, Popperian methodology lacks precise rules and reasons which show how its application allows us to reach our cognitive aims better than other methodologies. Nickles argues against both this extreme, which considers methodology a branch of logic and epistemology, and the naturalistic extreme that reduces methodology to an empirical science. Nickles insists on heuristic, which in Popper's methodology consists of very weak advice and in Lakatos' of detailed but epistemically irrelevant rules. For Nickles, heuristic plays an important role; it is a "theory of inquiry rather than of logic or epistemology in any narrow sense;" thus "which methods do work better must be determined by an extensive analysis of scientific practice, from study of human cognitive psychology and sociology, and from the economy of research".

Raimo Tuomela, takes a different approach, closer to Popper's. He defines science and scientific method by first pointing out certain essential components of a cognitive field K, and then establishing conditions on K. These conditions, considered "acceptable ideals" by Tuomela, work as a demarcation criterion between science and pseudoscience (which would differ from science with respect to every element of K) or protoscience (which would safeguard some elements or conditions of K). The examination of parapsychology allows Tuomela to exhibit how a pseudoscience violates this condition.

John Watkins has also developed an anti-naturalistic idea of the method of comparative valuation among theories. His proposal is articulated in different phases. First, he defines an "optimum aim" for science, (the Baconian/Cartesian ideal as he calls it) based on certain strong *desiderata* which may be divided into two components: (A) cognitive security and (B) depth of department. Then he tries to make this aim coherent and feasible by weakening A; the result is "the normative claim that, of all the competing scientific theories in a certain field that are possibly true in the above sense, the best is the one that is deepest and most unified, predictively powerful and exact". Finally he develops a technical instrument, a theory of corroboration, to effect comparisons among theories.

But prior to the specifics of comparing discrete theories and the detailed examination of case studies, Burian reminds us there is a deeper problem underlying considerations of rationality: conceptual change.

An adequate understanding of the justification for conceptual change is often held to be a prerequisite for developing a satisfactory account of the rational processes (if any) governing the development, evaluation, and choice of theories.

Despite such agreement, it is also the case that we do not have a unified terminology, agreed upon analytic tools, or a systematic characterization of conceptual systems to achieve the understanding we require to pursue our interests in rational theory change. Burian points to several confusions and misdirections in our talk about conceptual change. The central issues concern the extremism of historicist and autonomist views of the character of philosophy of science. He argues for restraints against emphasizing either position without consideration of the other, and demonstrates that a range of "competing values underlie the alternative interpretations of theories and conceptual schemes."

Do these various approaches provide us with a method of rational scientific change? No idea is free of difficulties. Further philosophical and historical analysis is necessary here as in every endeavor of this kind. However, the essays in this volume show that a lot of work has already been done and that the battle for the rationality of science has already scored many points.

JOSEPH C. PITT
Department of Philosophy
Virginia Polytechnic Institute
and State University

MARCELLO PERA
Dipartimento di Filosofia
Università degli Studi di Pisa

ACKNOWLEDGEMENTS FOR THE ITALIAN EDITION

All the contributions to this volume but one are original. They have been written for this occasion with the exception for that of Popper. The editors thank Sir Karl Popper for making available his essay and all the authors for having interpreted the goal of his volume so well and for having accepted the role requested of them in the general plan of the book.

ACKNOWLEDGEMENTS FOR THE ENGLISH EDITION

The editors thank R. M. Burian for allowing his piece to appear here, even though it was originally written for another occasion (see Burian's "Author's Note"). We also thank Laudan, Nickles and Shea for their revised papers.

PART I

THEORETICAL CONSIDERATIONS
CONCERNING RATIONALITY AND
SCIENTIFIC CHANGE

RICHARD M. BURIAN

HOW NOT TO TALK ABOUT CONCEPTUAL CHANGE IN SCIENCE [1]

1. THE PROBLEM

Conceptual change in science has been a "hot" topic for more than two decades. Yet, in spite of the strenuous efforts of a good number of philosophers and historians, our understanding of conceptual change is very poor. In spite of general acknowledgement of the importance of conceptual change, there is no general agreement about what it is, how it works, how it should be evaluated, or how far-reaching its consequences are. There is considerable agreement that major changes of theory (which, for better or worse, Kuhn 1970 has taught us to call "scientific revolutions") are typically accompanied by some form of conceptual change. There is fairly general agreement that such conceptual change must be taken account of in examining the rationality of theory change and of the rationales offered by scientists for abandoning a theory or seeking to develop alternative theories. Conceptual change is often held to be of critical relevance to our understanding of progress in science and to our assessment of the ability of science to achieve such goals as maximizing truth and explanatory power, facilitating prediction or control, and achieving maximal simplicity of its theoretical foundations. An adequate understanding of the justification for conceptual change is often held to be a prerequisite for developing a satisfactory account of the rational processes (if any) governing the development, evaluation, and choice of theories. In short, conceptual change has become a topic of central importance in the philosophy of science.

In spite of intensive attention devoted to these issues, there is no settled consensus concerning the nature or the workings of conceptual change. There is no generally accepted terminology in terms of which to describe concepts, systems of concepts, changes in concepts, or changes of conceptual frameworks. There is no agreement regarding the analytical tools appropriate to the examination of conceptual change, the criteria for individuating concepts or conceptual systems, or the means of separating "mere" change of belief or change of theory from

3

J. C. Pitt and M. Pera (eds.), Rational Changes in Science, 3—33.

conceptual change. If there are such entities as conceptual frameworks, we do not have an adequate understanding of their interrelations with the languages employed in science, the research programs and theories employed by scientists, and the practices by means of which scientists give experiential or experimental content (and perhaps meaning?) to their theories.

This is not to say that our situation is hopeless. There are, after all, a number of competing stances regarding theory change. Among the more familiar, one might mention those of Feyerabend (1975), Holton (1973), Kuhn (1970, 1975), Lakatos (1978), Laudan (1977), McMullin (1979), Putnam (1975, 1978), Sellars (1968, 1973), Shapere (1974b, 1980), Stegmuller (1967), and Toulmin (1972). The sharp and fruitful controversies which have arisen from the development and interaction of these positions may well bring in their wake the resolution of many of the problems touched on below. Indeed, hoping that this is how matters will develop, I shall stand back from the fray in this paper, proceeding in relative independence of any particular stance. I shall suggest some constraints, or criteria of adequacy, which an adequate account of conceptual change ought to satisfy. In addition, I shall try to show, first, that the intractability of certain of the problems that have been worrying philosophers is, in part, a consequence of the unsuitability of their analytical tools and, second, that it is desirable to shift the focus of debate so as to incorporate an examination of the values underlying philosophical analyses of science.

2. THE RELEVANCE OF VALUES

Behind most of the difficulties which philosophers have encountered, there lurks a confusion of fact and value. To combat this confusion, I advocate explicit attention to first and second order issues regarding the *value* of different sorts of explanatory theories, research policies, methodologies, standards of acceptance, and so on. At the first order, there has been a great deal of debate about how science works, about what the entities (theories, explanations, methods, conceptual frameworks, etc.) encountered in science *are*. Much of this debate has dealt with value-laden questions in a disguised and confused way. The issues at stake include: What should science be? What would be the most fruitful way of looking at the general features or the structure of science and of scientific reasoning? In what circumstances, if any, should we prefer scientific to prescientific or extrascientific knowledge? Should we draw (cognitive) limits around science? Should we treat it as the ultimate arbiter of all factual questions? How do we evaluate changes of value within science? Is conceptual change justified? And so on. At the second order, the questions become more difficult. How should we evaluate and what should we count as adequate support for such claims as the following? Conceptual scheme A, theory B, and research policy C are of greater cognitive value than their competitors. Again: Theories are better understood as changing components of research programs rather than as sets of statements. Yet again: A particular transition is better understood as a change of conceptual system than as a "mere" change of theory.

Three benefits, I suggest, derive from placing such explicitly evaluative questions at the center of the debates over conceptual change: (1) It will eliminate many confusions which have arisen from running descriptive and evaluative issues together. For example, in spite of the practice of some "historicist" philosophers, resolution of the *descriptive* question, "How do scientists (how does a particular scientific community) choose between competing theories?" makes no *direct* contribution to solution of the *evaluative* question, "How should scientists choose between competing theories?" Such confusions between descriptive and evaluative questions are very common in the contemporary debates. (Cf. Musgrave 1983.) (2) It will reduce the temptation to seek unique answers, to suppose that there is only one valuable policy, only one "correct" analysis of theories and so on. There are many kinds of cognitive values (including long and short term values) and there are many purposes germane to the evaluation and analysis of science. While it is often necessary to balance these against each other, while it is sometimes necessary to choose one way of working which *on balance* seems best, it is by no means always necessary or advisable to use the same evaluation, system of evaluation, or analytical apparatus for different purposes or when dealing with different values. (3) By making explicit the various values which are at stake, the turn to specifically evaluative questions will force philosophers to address a central methodological difficulty that has infected much of the literature. The difficulty concerns the relation of different values to each other and of facts to values. Is there *any* role for factual evidence in the evaluation of different philosophical positions regarding what science *should* be?[2] If so, what is that role? What balance of values should be employed in evaluating scientific work? And how should one balance values which, in application to a particular case, turn out to compete with each other? (I have in mind such frequently conflicting first order values as simplicity and exact fit with experimental results and such second order values as fit with abstract theories of knowledge and detailed applicability to case studies.) Among the difficulties of this general sort are those concerning the degree and source of the authority of the values propounded by philosophers. Such difficulties are likely to preoccupy philosophers of science throughout the 1980s.

3. AUTONOMISM AND HISTORICISM

Perhaps the best way to set out the issues which concern me is to examine two extreme positions which function as the Scylla and Charybdis between which most philosophers working on these issues seek to thread their way. My portrayal of these extremes is not intended to reflect the views of any individual. Rather, I present two philosophical positions as ideal types, each very tempting and each untenable in its extreme form. Most workers in the field fall somewhere between the two extremes.

At one pole is the view that (general) philosophy of science is an autonomous discipline, that there is a first philosophy or a philosophical methodology from which, by some sort of analytic, *a priori*, or transcendental techniques philosophers can derive, dictate, or legitimate the standards by which scientific work should be evaluated epistemologically, methodologically, or ontologically. The *application* of such standards to particular explanations, hypotheses, or theories requires, of course, substantive knowledge of science, but one is supposed to be able to derive, establish, or justify the standards independently of particular scientific knowledge; *establishing the standards is a philosophical rather than a scientific enterprise.* I shall call such philosophies "autonomist."

At the other pole are "historicist" philosophies of science which claim, with the later Wittgenstein, that philosophy leaves everything as it is. According to the historicists, it is not the philosopher's business to derive, dictate, or legitimate the standards by which scientific work should be judged. Yet there is plenty for philosophers to do. For example, they might (1) articulate explicitly and perspicuously the standards employed by actual scientists (a task which even the best scientists often find difficult), (2) disentangle the different historical and disciplinary contexts in which different standards "are appropriate,"[3] or (3) clear away the confusions which result when the standards "appropriate" to one context are applied out of context — e.g., by historians or philosophers who seek to judge past theories by present standards.

Logical positivism and various logical empiricisms provide relevant examples of philosophies close to the autonomist pole. In virtue of *a priori* considerations or of some form of analysis, these philosophies treat science as *defined* by some such goal as maximizing justified true belief about the world or maximizing the scope and simplicity of our experientially grounded knowledge. Accordingly, what counts as the course of historical science is, approximately, the accumulation of "hard" facts and of ever more exact, comprehensive, and systematically unified theories covering those facts.[4] On general philosophical grounds, most positivists and empiricists held that (in principle) the sum total of scientific knowledge at any given time can be stated in a "unified" language consisting of a theory-neutral ("observational," "fact-stating") base and various theoretical sub-languages built on that base.

Given this account of science, all the important, philosophically explicable developments in its history conform at least approximately to philosophical accounts of scientific rationality — if for no other reason, then because for something to *count* as science (or as doing science) it must aim at or yield additions to the base of factual knowledge or to the superstructure of comprehensive explanatory theories. Accordingly, so-called "internal" historiography is (in principle) adequate for understanding how the achievements of historical science were justified even though there are phenomena (such as the discoveries and inventions of creative geniuses) for which no philosophical understanding is possible.

In the autonomist tradition, much of the philosophy of science is simply applied epistemology. The standards which should be employed in judging the adequacy and truth of factual claims, the degree of support for hypotheses and theories, and the justification for and value of theories, are prior to the truths of science; they derive from philosophical theories of knowledge, justification, and so on.[5] The status of *particular* explanations, hypotheses, and theories depends on scientifically established facts as well as on these standards. The primary philosophical tool with which to work out such matters is logical analysis of the (idealized) language of science. Use of this tool allows one to set forth the structure of (sound) confirmations, explanations, and theories and to compare "rationally reconstructed" explanations, theories, and so on against the resultant norms. Such "rational reconstructions" — full, regimented statements in philosophically perspicuous language of the data, hypotheses, explanations, and theories, in short of the "finished products" of science as of a particular stage in its development — are the proper object of philosophical investigation since they provide the most perspicuous and advantageous account of the materials to be evaluated.

A number of recent philosophers — Feyerabend, Kuhn, and Toulmin can serve as examples — fall close to the historicist pole. Formally speaking, they may allow science to pursue some overarching objective such as maximizing justified belief about the world. Nonetheless, each of these philosophers treats the *effective* goals and standards of judgment in science as varying with time, as shifting with larger traditions and major theories. They tend to see the course of scientific history as involving two quite different phases or moments. One phase, labelled "normal science" by Kuhn, involves self-corrective accumulation of facts and development of theories and explanations which have already been sketched out. The other phase — Kuhn's "scientific revolutions" — is characterized by changes of "world-view," "paradigm," "highest-level theory," or "explanatory ideals." In the latter phase, there can be *loss* of facts and explanatory power, change of standards governing acceptance for fact-claims and for theories, and loss of the ability to compare rival theories in a straightforward way. These difficulties arise because of conceptual change. Such change undermines the hope of working in a "theory-neutral" language, unifiable to cover the whole body of science. In order for observations to succeed in testing a theory, they must be reported in a language compatible with the categories of that theory. But since differing fundamental theories employ incompatible or incommensurable categories, the languages in which observations germane to those theories are reported cannot be adequately mapped onto each other. Or so the more extreme historicists would have us believe.

Accordingly, because of the radical shifts in the conceptual systems of science imposed by theory change, no attempt to account for the rationality of central revolutionary episodes in the history of science can be entirely successful: to overthrow theories on the basis of evidence, one must employ evidence whose relevance or power ought *not* be admitted by the supporters of the overthrown theories. Accordingly, *in principle*, a proper understanding of the course of scientific history requires the employment of a (partly) "external" (e.g., sociological) historiography to resolve why one theory (or community) won out over another, and why it did so when it did.

In this tradition, philosophy of science can never be the "mere" application of epistemology, for *what counts as adequate means of supporting hypotheses is something learned, in part, from the study of science and its history.* Accordingly, whatever standards independent of science are put forward by epistemologists, they are too weak to accomplish the evaluation of scientific claims or theories unless they are further specified or revised *on the basis of the relevant science.* The source of authoritative standards is, in good part at least, internal to the "paradigms," "explanatory ideals," and so on found in science; such standards cannot be judged by "pure" philosophy. Normatively, the most the philosopher should aim at is to set forth and apply the rational standards for acceptance, development, correction, and improvement of ongoing science *implicit within science itself.* The primary philosophical tool with which to pursue such a task is historical analysis of the conceptual frameworks, theories, explanations, norms, and practices of historical science. Such analysis is unreliable if it is restricted to the "finished products" of science (let alone rational reconstructions thereof) since these, in their attempt to conform to a traditional and narrow canon of evidential support, notoriously obscure major aspects of the development and evaluation of scientific work.[6]

4. DIFFICULTIES

Anglo-American philosophy of science, which was predominantly autonomist twenty years ago, appears headed toward an historicist phase. This purely descriptive claim ought not, of course, count toward the evaluation of the historicist and autonomist stances; perhaps historicism will prove to be only a fad. What matters for evaluative purposes are the considerations and arguments affecting the balance between the two positions. A full attempt at assessing this balance goes far beyond the confines of a single paper. On this occasion, I shall simply present three difficulties at each end of the spectrum, beginning with autonomism. I shall maintain, perhaps a bit dogmatically, that these difficulties undermine the application of both doctrines in their pure forms. The discussion of the difficulties involved is important even if this claim proves exaggerated; each difficulty suggests a constraint which, I believe, ought to be satisfied by any theory of conceptual change concerned to evaluate, or provide standards applicable to, historical case studies.

5. DIFFICULTIES FOR AUTONOMISM

1. Historians of science have pointed to a great variety of more-or-less abrupt ruptures, breaks, shifts, and discontinuities in the histories of particular sciences. Among the features of various sciences which have been claimed to exhibit such discontinuities are the language employed (including observation terms, theoretical terms, rules of inference, and objects referred to), the concepts employed, the descriptions of data, the interpretation and weighting of experimental results, the range of data and experimental results considered relevant, the modes of perceptual experience of the scientist, the agenda of problems to which the scientist is held responsible, the fundamental theory accepted in a given field, the methodological practice in that field, the evaluative practices by which scientists judge fact-claims and theories, the official methodological and epistemological standards by which work in the field is (supposed to be) judged, and the metaphysics or ontology or world-view underlying ongoing work in the field. Obviously, these factors are not all independent. Obviously, not every alleged discontinuity is genuine. But there is a great deal of smoke here — and, I am convinced, more than a bit of fire. Indeed, I think it fair to summarize the results in question as follows: historical scientific communities typically reevaluate facts and theories in the light of revised evaluative standards as well as new information.[7] These standards, in turn, are revised in the light of what that community — or what wider scientific communities — learn about the world in the course of investigating various phenomena of concern.[8] If this is correct then an autonomist philosophy faces three alternatives. (1) It must explain the revision of standards in question as mere *specification* (application to particular cases) of more general standards which the philosopher provides (or could provide) *a priori*, or (2) it must deny that *real* changes of standard take place, or (3) it must deny scientific status to all work except that which conforms to, or employs, some one favored standard.

This difficulty suggests the first of the constraints which I shall propose: *any adequate theory of science ought to provide, in scheme at least, a means of accounting for and evaluating* (if only to discount) *the discontinuities just mentioned.* The scheme should be usefully applicable, without undue historical distortion, to some of the more interesting cases of discontinuity which are known. (I will allude to one of these in discussing the next difficulty.)

2. Two forms of discontinuity, which Doppelt (1978) has labelled "loss-of-data" and "shift-in-standards" are of special interest.[9] The first is loss of the ability to account for a class of facts; the second includes as a special case abandonment (as a constraint on acceptable theories in a certain field) of the requirement that those theories be able to account for or explain a certain class of facts. Consider one of the central examples in *The Structure of Scientific Revolutions*, as reviewed and sharpened by Doppelt.[10] At stake is the transition from pre-Lavoisierian "qualitative" chemistry, in which the theories of elective affinity and phlogiston played a central role, to Daltonian "quantitative" chemistry. Of interest to us is a large class of facts which were accepted by both qualitative and quantitative chemists. Among these are the qualitative similarities among metals (which are far more like each other in specific, readily identified ways than are the ores or "earths" from which they are extracted) and the parallel reactions, for example in the formation of acids, manifested by a number of metals. According to some authorities, a major accomplishment of qualitative chemistry was its account of these facts. The common properties of metals were accounted for by the (supposed) fact that all metals contain an excess of phlogiston, which they readily give off. ("Phlogiston" was the label for an acidic principle which was also thought to be the matter of heat.) Metal-bearing ores, in contrast, were supposed to be disparate, unrelated substances, generally low in phlogiston or manifesting a high degree of affinity for that substance. Historians' treatments of this case make it clear that many qualitative chemists held that their account of the common properties of metals set a standard of adequacy for general chemical theories: any adequate chemical theory should account *at least as well as theirs* for the relevant properties.[11]

Neither Lavoisier's nor Dalton's chemistry offered any account of the similarities among metals or of the reasons for which they entered into so many similar chemical reactions. Nor did either chemistry explain these matters away by denying or correcting the facts in question or by making them the responsibility of some other science. In effect, the new chemists simply deemphasized these facts or abandoned all concern with them. Effectively, they ignored some of the central accomplishments claimed by the preceding chemical theory.

This is not to deny that phlogistic theory was in deep trouble. It failed utterly to account for the weight relations in chemical reactions (the account of which was the crowning achievement of quantitative chemistry) and its efforts to cope with this problem led to internal difficulties so serious that the whole theory was called into question and eventually abandoned. Nor is this to deny that the resultant difficulties raised questions about the *value* of the phlogistic account of the qualities of metals;[12] if there is no such substance as phlogiston, the similarities among metals cannot be explained by their high phlogiston content. But *why should the inability to account for the qualitative properties of metals count less against quantitative chemistry than the inability to account for weights counted against qualitative chemistry?* How does one — can one — justify the shift-in-standards which allowed Lavoisier's and Dalton's chemistries to bypass this touchstone? This sort of dilemma is faced repeatedly in the history of science.[13]

The point is not tied to the particular case. Whether or not one accepts Kuhn's assessment (1970, p. 107) that in this instance the new theories "ended by depriving chemistry of some actual and much potential explanatory power," the fact must be faced that, at least in the short term,[14] scientific progress is very often purchased at the price of a loss of explanatory breadth and power. That is, scientific "progress" is often achieved by ignoring, or dismissing as irrelevant to the evaluation of current theories, facts which ought to be accounted for according to previously prevalent standards not *specifically* shown wrong. These considerations suggest our second constraint: *evaluative philosophies applicable to case studies ought to be able to provide a rationale for, or an evaluation of, the sort of shifts-in-standards just illustrated.* That is, they should provide grounds that allow rational pursuit and acceptance of theories which have failed (so far) to explain a relevant range of phenomena even when those phenomena are not "farmed out" to another theory. They should be able to explain how, in the absence of "direct" new evidence, epistemological reevaluation of theoretical approaches earlier thought not worth pursuing can convert those approaches into "serious possibilities." What is required is an answer to the question, "In what circumstances are shifts-in-standard of the sort illustrated reasonable?" If this question cannot be answered, many, perhaps virtually all, large-scale theory transitions of the sort examined by Kuhn may have to be judged rationally unacceptable.

3. Typically, in the early phases of the development of a major theory (and sometimes long after), there is an immense amount of material which it is clear that the theory *ought* to be able to account for, but which it has not yet been able to handle. Sometimes the theory itself is not worked out well enough; sometimes our knowledge of auxiliary hypotheses, auxiliary theories, and boundary conditions is inadequate. Very often the theory itself suggests strategies for attempting to resolve the difficulty, but frequently it simply is not clear what is wrong. Let me mention three stock examples. (a) During its first one hundred years, the Copernican theory of the heavens was confronted with a series of paradoxes which, according to Aristotelian physics, made it ridiculous to suppose that the earth is moving. (On Aristotelian principles and available data, one could deduce that if the earth were rotating, there would be a roughly thousand mile an hour counterwind. As is well known, no such wind is observed.) But, speaking very crudely, we now realize that it was Aristotelian dynamics and not Copernicanism that was at fault, that the evidential worth of the lack of a counterwind is altered by the overthrow of Aristotelian physics. (b) During the first one hundred years or so of Newtonian mechanics, no one was able to obtain a Newtonian proof that the orbits of the planets are stable. This elementary, but very difficult problem was resolved by LaPlace's perturbation theory. (c) Darwin's theory of natural selection is unable to account in detail for the evolutionary history of *any* population or species or biota in the absence of an adequate genetic theory. Darwin's *Origin* implicitly required, but did not offer a detailed genetics. It had to be judged in the absence of an adequate genetics. And his later theory of pangenesis was clearly inadequate; indeed, if one took it seriously, it raised far more difficulties for natural selection than it resolved.

The point of these examples is this: almost always, what scientists are interested in evaluating is not a closed system or a finished theory, especially not a theory regimented along the lines proposed by philosophers of science. (Even now, the neo-Darwinian theory of evolution is not a closed system and has not been put in a thoroughly regimented form.[15]) Rather, scientists must deal with theories which are overlapping open systems,[16] all of which face at least some difficulties. Their assessments of theories and programs of research, evaluations of explanations, weighing of auxiliary hypotheses, and so on, typically occur in an "open" context. In such a context, the lines separating disciplines, the lines determining which evidence is relevant and which is not, cannot be taken for granted. In such a context, evaluation of theories and explanations is linked to practical questions regarding how to proceed when adequate evidence (adequate, that is, by an abstract philosophical standard) is not in hand. The relevant issues concern how one ought to proceed when one is uncertain about how rigorous the appropriate standards of acceptance ought to be and how to allocate the blame for the unresolved difficulties among the various relevant theories and practices.

The point is *not* that an autonomist evaluation of a deductively closed body of statements cannot be carried out. In principle one can fix the body of propositions that one will include in each relevant theory at any moment, regiment the resultant theories appropriately, and evaluate them by whichever autonomist standard. The point is, rather, that once one has executed this formidable task, the relationship of the resultant evaluation to the evaluations and choices which scientists ought to make is highly problematic, especially since the *purposes* of philosophical and scientific evaluation differ. (Cf. Burian 1977, esp. pp. 12 ff.) To a certain extent, the evaluative apparatus employed by the philosopher freezes the evidential relations and the scientific materials involved. (Given the theories available in 1575, the lack of a counterwind *should* have been assessed as timelessly valid evidence against the rotation of the earth!) If the philosopher wishes to relate his work to ongoing science, it is incumbent on him to relate his evaluations (which may well work very well for retrospective evaluation of completed research programs (Doppelt 1978, sections VI and VII)) to the short-term evaluations which the scientist, in the middle of things, must make in the course of his work. And, as Nickles (1980) has recently argued, simple recourse to the traditional distinction between the context of discovery and the context of justification is of no use in this connection.

These considerations yield our third constraint: *theories of conceptual change ought to examine and evaluate the considerations relevant to* scientist's *evaluations of ongoing work.* Accordingly, if they employ autonomist techniques (which treat the materials the scientist deals with as closed systems), *they should appraise the relevance of the resultant evaluations to the evaluations which the scientist must make* in mediis rebus. To the extent that pure autonomism treats the materials in its rational reconstructions as finished products, it will find it especially hard to carry out such appraisals and especially hard to meet this constraint.

6. DIFFICULTIES FOR HISTORICISM

It is time to balance the account by raising three difficulties for historicism.

1. In its pure form, historicism describes but does not evaluate the standards which (it claims) are found in various conceptual schemes, explanatory ideals, paradigms, research traditions, or world views. (I use all these labels only to let the historicist have whatever unit of discourse he chooses.) Historicists consider it just as anachronistic and mistaken to evaluate previous theories or explanations by current standards of acceptability or pursuitworthiness as it would be to claim that their proponents should have rejected them simply because, as we now know, they are false. Historicists hold that each major scientific tradition sets its own problems and its own standards for their solution. They maintain that since these "internal" standards are anchored by the problems set by the tradition, they are the ones in terms of which the ongoing work in the tradition should be evaluated; a shift-in-standards amounts to a shift in problems and results in misevaluations of work (or results) addressed to the old problems.

Acceptance of this extreme position poses a serious dilemma. Either the historicist is left with an indefensible normative position (which I will criticize in the next paragraph) or with a *soley descriptive task* — to wit, describing the traditions of concern and the work performed in them in such a way that one can determine *as a matter of fact* which work conforms to the (avowed or implicit) standards of which traditions. It should be clear that the descriptivist version of historicism abandons *all* independent evaluation: its proponents cannot coherently answer such questions as "Was Darwin's theory well supported?", "Should it have been accepted (or accepted in England as of 1880)?",[17] or "At what point was it reasonable to switch from Aristotelian to Galilean dynamics (or from an Aristotelian to a Galilean conceptual scheme)?" If it can be established that a particular tradition would have answered these questions in a particular way, this historical result fails to answer, indeed evades, the evaluative questions just posed, for the historicist has no standpoint from which to evaluate the different answers offered by different traditions.

This last point helps show that the normative version of extreme historicism is even less appealing. By refusing to allow any "external" (that is, tradition-foreign) standard to be applied to the norms of (or work within) a tradition, each tradition is accepted as the arbiter of what, for a certain period or for a certain disciplinary context, counts as good science. Disagreements between traditions *automatically* degenerate into a right-for-me wrong-for-you impasse. The result is an abdication of any serious role for philosophy. But the uncritical acceptance of each tradition as self-certifying, the refusal to allow external evaluation of such traditions or their norms is an indefensible normative policy.

Whichever version of extreme historicism one considers, these objections suggest our fourth constraint: *an adequate philosophical theory of conceptual change ought to enable serious evaluation of the explanations, norms, theories, traditions and* (if such there be) *conceptual systems of science.* Putting the point more concretely: one's account of conceptual change ought not exclude, *a priori*, arguments which show that a whole tradition has gone awry. For example, it must make sense to argue from an external point of view that a particular research tradition is likely to lead away from the truth, that it has substituted *Scheinprobleme* for genuine problems, that it employs inappropriate, unfruitful, or mistaken standards of acceptability, pursuitworthiness, and so on, that it consistently yields mistaken evaluations of research policies and theories, or, perhaps, that it ought not even be counted as a *scientific* tradition.

2. The second difficulty concerns the holism and relativism which result from extreme forms of historicism. According to some extreme historicists, the contextual restrictions on the standards by which scientific claims (methods, theories, and so on) are evaluated arise because every claim one makes is infected with one's theory, paradigm, or world-view. Thus when a Copernican says, "Sunrise took place today at 7:37 AM," in spite of the etymology of the phoneme "sunrise," the claim he makes could not be accepted by a Ptolemaic. The Copernican, after all, is claiming that the earth rotated so as to reveal the sun at 7:37, while the Ptolemaic understands the same sounds as claiming that the sun literally rose above the horizon. The consequence of such a doctrine is that major conceptual schemes can be compared, if at all,

only as wholes, not "locally". For if we cannot rid any level of our discourse of its commitment to our world-view, then even when we make a specific prediction which *seems* to disagree with one made by our opponent, we will be talking at cross-purposes. On such an account, if the Ptolemaic predicts that sunrise will occur at 7:38 and the Copernican predicts that sunrise will occur at 7:37, their predictions are *not* contrary but, to use a term from Kuhn and Feyerabend, "incommensurable".[18]

Incommensurability works as follows. Suppose that the orb of the sun clears the horizon at 7:37. Because the time disagrees with the Ptolemaic prediction, the Ptolemaic has an *internal* problem — he needs to make adjustments somewhere in his corpus of beliefs to account for his false prediction. But because the sun — literally — rose for him, the fact that it rose at 7:37 rather than 7:38 in no way supports the Copernican's account of the matter. To use Kuhn's dangerous and mistaken metaphor, the Ptolemaic and the Copernican "live in different worlds."[19]

In real life, of course, we do not have incommensurability of this kind; we live in the same world, we refer to the same sun, and when, as in a case like this, we talk past each other, we do not do so totally. The Ptolemaic and the Copernican typically manage to describe the sunrise in a way which is neutral as between their theories. (Such a description will still contain a variety of presuppositions — e.g., that the sun is an enduring body — and thus claim more than is, in the narrowest sense, justified by observation alone.) Their observations of the sunrise can then be used as a means of determining that, *other things being equal*, this particular Copernican prediction yields some support for the Copernican theory and some disconfirmation of the Ptolemaic theory of the behavior of *our* sun, of the heavenly bodies in *our* sky. Local comparison between rival theories, even when they are revolutionary alternatives, is something we manage to accomplish, at least on occasion.[20] Indeed, for a variety of reasons too complex to develop on this occasion, I think that local comparison is normally possible between contemporaneous versions of competing theories when these are sufficiently developed to yield significant predictions.[21] Accordingly, we reach our fifth constraint: *an adequate theory of conceptual change must not exclude*, a priori, *point-by-point comparison between rival versions of rival theories.* That is, it must allow us to identify problems, such as the timing of sunrise, which the theories share and allow us, in principle, to determine which version, if any, handles those problems better than its rivals do.[22]

3. The third difficulty for historicism concerns a particular part of the argument against local comparison of theory versions based on rival world-views — namely, the historicist theory of meaning. A key step in the denial that a Ptolemaic and a Copernican could put forward contrary claims about the sunrise was the claim that each must mean something different by the term "sunrise" because, as each scientist used it, the term incorporated or presupposed the laws or principles of his theory. Thus the meanings of *all* terms relevant to the testing of a theory presuppose the principles of the theory being tested, and the meanings of terms built on different presuppositional bases are incommensurable. (Where no explicit theory is the source of the relevant presuppositions, such presuppositions still abound. They can be drawn from one's world-view or from the commonsense commitments of one's cultural circle.)

As this argument stands, it cannot be clearly evaluated because of the obscurity of the notions of *meaning* and *presupposition* to which it appeals. For present purposes, however, a fairly crude way of approaching the matter suffices. The criticism proposed here is based loosely on the position taken by Putnam some twenty years ago in "How not to Talk About Meaning." Putnam argues that we need not "discard ordinary language" even when it "presupposes false beliefs which we wish to reject in our pursuit of truth" (Putnam 1975, Vol. 2, p. 126). We can change many of our commonsense beliefs without so altering the meanings of the terms or claims of ordinary French, German, English, or Hopi that we have to change languages altogether. This point, I suggest, carries over to scientific languages: we can change central theoretical commitments about, say, the motion of the heavens or about the quantization of momentum without so altering the use of such terms as "sun" and "sunrise" or "electron" and "proton" as to have to change languages altogether. That is, in spite of the fairly radical changes of belief and the differences of connotation or sense (if we can make these notions precise) which accompany theory change, we can still employ a common language to reidentify some of the objects or situations in question and to localize certain occasionally testable differences of expectation about those objects or situations.

I am myself fairly convinced of the Quinean view that considerations of meaning are altogether useless here. However, even if all philosophical doctrines of meaning must, in the end, be rejected as mistaken, we can still distinguish between those which are worthy of serious consideration and those which are not. Any theory of meaning which *automatically* excludes point-by-point comparison between rival theories (or theory versions) belonging to competing scientific traditions (paradigms, etc.) ought to be rejected out of hand. It may be that we are locked into incommensurable languages, though I doubt it, but it is not *a priori* true that we are caught in such a bind, it is not true simply on the basis of the way we give meaning to our terms and claims. That is, our final constraint governing *an adequate theory of conceptual change* is that it *ought not be based on, ought not entail, presuppose, or otherwise require, a theory of meaning according to which we are inevitably trapped in incommensurable languages when we have fundamental theoretical disagreements.*

7. CONSTRAINTS SUMMARIZED

A summary of the six constraints I have propounded is in order. (1) An acceptable philosophical account of the phenomena of conceptual change ought to contain apparatus facilitating evaluation of the discontinuities exhibited by historical science — discontinuities in fundamental beliefs, evaluative standards and methodologies, systems of concepts, theories, and so on. (2) Specifically, it should be prepared to evaluate those shifts-in-standard which, at least in the short run, permit reductions in the explanatory scope and competence of acceptable theories (what Laudan has somewhere called "Kuhn-loss"). (3) These evaluations should not be applicable only retrospectively; they should also be applicable to ongoing work in which the precise boundaries between theories and disciplines, and the precise content of particular theories, is not settled. (4) The philosopher's evaluative apparatus should not exclude, *a priori*, comparative evaluations (which are, necessarily, at least in part from an "external" point of view) of the rival "conceptual schemes," "explanatory ideals," "paradigms," and so on pertinent to theoretical science.[23] (5) More specifically, it should not automatically exclude point-by-point comparison of the predictions of contemporaneous versions of rival theories belonging to rival research traditions. Finally, (6) if a theory of meaning is employed by the philosopher or built into his apparatus, it should not, by itself, exclude such point-by-point comparison.

8. A GLANCE AT TOOLS

There are two further matters which I wish to address briefly. The first is the adequacy of the analytical tools philosophers employ, the second is the desirability of placing evaluative questions at the center of the debate.

It is well known that various kinds of language analyses, both formal and informal, are among the major tools of contemporary philosophy. Not surprisingly, this is evident in philosophy of science; much important work has been done by means of (mostly formal) analyses of the language(s) of science and of different theories and by means of linguistic analyses aimed at revealing the meaning or the reference of scientific terms, and so on. These tools are very valuable and some of them may prove virtually indispensable for the clarification and resolution of important philosophical questions. Nonetheless, I suggest that the analysis of scientific languages, both natural and artificial ones, cannot serve as *the* means for analyzing or appraising conceptual systems or conceptual change. For instance, the question whether one is forced to some form of (conceptual scheme) relativism is *not* resolved by showing that we cannot isolate some theory-neutral or conceptual-scheme neutral language or sub-language within which to state and test the consequences of competing theories, within which to analyze competing conceptual schemes. Meaning variance, unless it is total, does not entail relativism and ought not, by itself, make relativism plausible (Leplin 1979). Without going into detail, I shall mention some limitations of language analysis as a tool and some issues regarding conceptual change which it is not suited to resolve.

How do languages reflect one's world-view, one's fundamental pre-suppositions, one's concepts?[24] There is no neat connection between an individual's or a community's system of concepts or world-view and the (natural) language it speaks. It inherits a language that bears traces of older views and it leaves behind a slightly altered language. The expressive power of natural languages is not fixed. Indeed, one may plausibly hold that when one allows for the dynamics of languages, when one allows the speakers of a language to learn new theories, when one allows speakers to add various technical extensions to their languages, any theory or any world-view can be expressed, however awkwardly, in any natural language. (See Levin 1979). This does *not* mean that language is unimportant, that notation is unimportant, that there is no connection between fundamental beliefs and mode of expression, or that linguistic reform is irrelevant to the development and evaluation of theories or systems of concepts. It does not mean that a community's commitment to such categories or concepts as *substance* and *accident* leaves no traces on its language. On the contrary. Nor does it mean that one can explain quantum mechanics easily in ordinary English or the calculus easily in Latin. Nonetheless, Latin *does* have adequate expressive power for doing algebra or calculus. The misfit between Roman numerals and algebraic manipulation is not a matter of the conceptual content or structure of Latin; the inherit limitations of Latin notation do not prevent proofs of theorems of the calculus even if, as a practical matter, the calculations are too complex for us to manage without alternative conventions. In general, it *is* possible to find a way of stating the fundamental claims of molecular genetics and Freudian psychology, quantum physics and Aristotelian mechanics within extended versions of Demotic Greek or Hindi, Hopi or English, Swahili or Thai. In short, different natural languages may be more or less suited to the expression of different theories and different world views, but their expressive power is not fixed, their cognitive structures (if languages, as opposed to theories and world-views, can significantly be understood as having cognitive structures) may be extended and altered in ways that allow comprehension and comparison of alternatives of the kinds with which we are concerned. Should there be such entities as conceptual frameworks, analysis of the natural language(s) which a person or community employs is not powerful enough to reveal which conceptual framework(s) they are committed to.

What, then, of the specialized languages, sub-languages, and calculi employed by philosophers specifically to reflect the structure of theories or conceptual systems? Are these the proper tools for analyzing conceptual frameworks?

This is not the occasion for a full treatment of the issues raised by these questions. I shall simply touch on three grounds for supposing that there are limits — limits which it is important to understand — to the use of specialized calculi and formalizations for the purpose of delimiting theories and conceptual systems.

The first set of concerns is based on the technical limitations of the tools in question. Consider, for example, the matter of reaxiomatization. If one is after, say, the beliefs about the structure of the world built into or presupposed by such fundamental theories as Newtonian and quantum mechanics, one must face two facts. First, precise and adequate formalizations of such theories are extremely rare and extremely difficult to construct. We almost always have to do without. Second, once one has a precise formalization which is supposed to reflect "the" conceptual structure of the theory (or its metaphysical commitments), one can always produce alternative, substantially differing formalizations which have the same expressive power. That is, there are alternative formalizations of any specific, deductively closed formalized theory which, under appropriate transformations, yield the same total body of theorems on the basis of significantly different primitives and rules of inference. For example, as I understand it, Newtonian mechanics may be formulated using the notion of a field as a primitive or with no explicit term for fields, with the notion of force as a primitive or with no explicit term for forces, and so on. It is therefore not possible to determine which (if any) of the relevant concepts belong to "the" conceptual framework of Newtonian mechanics (or of a particular Newtonian community) solely on the basis of a consideration of an adequate axiomatization (or of all adequate axiomatizations) of some version or versions of Newton's theory.

A second technical concern grows out of a consideration of some of Hilary Putnam's recent work. Most of the technical languages employed in the philosophy of science are extensional. Putnam (1978, Chap. II) has shown that the formal structure of such languages is not powerful enough to enforce the standard ("classical") interpretation rather than alternative intuitionistic interpretations of the logical connectives. If his argument is valid (as I believe it is), when a theory is formalized in an extensional language, insofar as we are guided solely by the formalization we may legitimately interpret the theory in what Putnam calls a "quasi-intuitionistic" manner. Nothing in the formal treatment of the language (including the formalism of extensional semantics) rules out a *re*reading of the formal representation of, say, "there are electrons" as "there is a description D such that ⟨D is an electron⟩ is provable from the axioms of the formalized theory."[25] Such a quasi-intuitionistic interpretation allows a theory to determine its own reality; under this interpretation, acceptance of the theory's axioms is sufficient to determine which existence claims of the relevant formal language are true. In Putnam's words (1978, p. 29), on a quasi-intuitionistic reading, "'existence' becomes *intra-theoretic*."

Formal considerations cannot exclude quasi-intuitionism. It is, however, excluded by our fifth constraint; the application of a quasi-intuitionist interpretation to rival theories automatically excludes point-by-point comparison of the rivals, in violation of that constraint. Whether we should accept the constraint is, I grant, debatable. *But the debate will not be settled by formal considerations*. To that extent at least, formalization in extensional languages is not alone sufficient to resolve the debates about the effects of conceptual change.

One cannot, at least at present, expect to get happier results from formalizations in modal languages. Notoriously, the interpretation of available modal formalizations of scientific materials is fraught with even greater difficulties than is the interpretation of extensional formalizations. Whichever way one turns, even when all formal and evidential constraints on formalization have been met, significantly different readings of the formal representations of scientific materials remain open.[26] The analysis of the formal properties of formalized representations of scientific theories and conceptual frameworks is not powerful enough to fix "the" structure of the theories or frameworks which are in question.

Turning away from these technical matters, a more general consideration touched on earlier (pp. [9 & 14—15] above) provides a rather different kind of support for my contention that there is no uniquely correct or perspicuous analysis of *the* content or *the* concepts of a given theory or conceptual system. If, indeed, the theories which we wish to evaluate are typically open to revision and if, at least occasionally, the revisions in question yield new formal structures and new concepts, then when one "freezes" the formal structure via a rational reconstruction, one specifies an object of investigation whose relation to "the" theory or conceptual scheme under investigation is problematic.[27] Precisely this problematic relationship is not susceptible to investigation by formal techniques.

The third consideration regarding the limitations of formalization as a tool introduces the final topic of this paper. I maintain that what we ought to count as the proper representation of a theory, conceptual scheme, and so on (or what we ought to count as the proper interpretation of a formalization) depends on our purposes and our values. If this contention is correct, it grounds an argument for the value of placing values at the center of the ongoing debate about conceptual change.

9. VALUES AGAIN

Throughout this paper, I have written naively as though it makes good sense to suppose that there could be a unique target of analysis — *the* concept of a conceptual framework, of an explanation (or explanation of a certain kind), of a theory (or theory of a certain kind), and so on. It is time to take account of the fact that this reification is mistaken and misleading. (Burian 1980, esp. pp. 318—319 and 328 ff.).

I have maintained that the content and structure of theories changes with time. So, too, does our concept (or, if you prefer, our conception) of explanation and justification, of theories and conceptual frameworks. Correlated with these changes are others touching on our understanding of what it is to provide an adequate account of a domain of phenomena, to provide support for theoretical hypotheses, to understand the phenomena with which we are concerned. It is extremely useful and important to give "hard-edged" reconstructions of these matters, i.e., to give precise metatheoretical explications of acceptability or of degree of support, of scientific understanding, of "the" structure of conceptual schemes, explanations, and theories. Unless we do so, our disagreements will be unfocused, our debates will be mere clashes of differing intuitions, we will not realize when vagueness has led to error, and we will not be able to develop trenchant criticisms of erroneous views. But the hard-edged analyses with which we work are best viewed as *tools*, as devices for sharpening our *proposals* concerning what we should make of the explanations, justifications, theories, and so on offered by scientists — and also concerning how we should improve our ways of explaining, justifying, and theorizing. As proposals, our analyses must be evaluated — and how we evaluate them depends on our purposes, on the values we are seeking to realize, and on our estimate of the way the world is.

The role of this last factor should not be underestimated. We learn how to learn in the course of learning about the world (Shapere 1980). Our account of science is influenced, as it ought to be, by what we think the world is like, by the kinds of science we consider successful. For example, it might have turned out that successful theories were all "phenomenal" and that "compositional" or "deep-structure" theories never yielded greater understanding or solved more empirical problems than the phenomenal theories with which they are correlated. Were the world like this, our accounts of scientific understanding would be vastly different than they are; we would value very different features of science than the ones we do as contributing to scientific understanding of the world. (Cf. also Shapere 1974a.)

Such a shift in values would alter our evaluation of "rational reconstructions" (which are proposals concerning the proper understanding) of the structure and content of conceptual frameworks, explanations, justifications, and theories, both in general and in particular cases. This illustrates the general point I am after: the values we hold, the features and items which we value highly, enter (and ought to enter) into our evaluation of rational reconstructions and thus affect which features of conceptual schemes, explanations, justifications, theories, and so on, we treat as "essential"; in short, our values do and should affect what we take such entities as theories to be. Since values have this effect, it is wise to recognize their role explicitly and to bring some account of the values which different explications serve into the debate over the explication of metascientific terms.

The recognition that the acceptability of a particular rational reconstruction can be altered by what we seek from science, by what we value in it, and by the purpose of our investigation, serves as a salutory reminder that the units of discourse in philosophy of science are not constrained to fit some Platonic form of explanation or of theoryhood, and so on. There is no "given" structure that the units of philosophical analysis must match. Rather, we choose and invent our units of discourse in order to make sense of science, which is a human enterprise, shaped in part by our image of it, and reshaped when that image changes.

I hasten to add that we are not free to accept any old account of science or of such entities as theories. Although we are free to choose the units of discourse which we believe will enable us to determine or appreciate the features of the objects of concern which give them the characteristics we value, if we choose our units wrongly, we will fail to accomplish our purpose. Nor are we free to choose our values whimsically. Science is an ongoing enterprise which we cannot reshape at will. So we must recognize that our strategies of analysis and our choice of units of discourse may be poor and that our values may be foolish, internally incoherent, "other-wordly" (and hence irrelevant to the evaluation of the merits of scientific knowledge in *this* world), or otherwise ill-suited to an understanding or evaluation of the scientific enterprise. But none of this alters the fact that our values play a crucial and largely unappreciated role in our determination of what science is and in our acceptance of some particular analysis of such targets of analysis as *conceptual change*. Debates about the proper analysis of such targets that do not take this fact into account are likely to miss the mark.

To make the point a bit more concrete, consider the use of our fifth constraint ("Do not automatically exclude point-by-point comparison of rival versions of rival theories!") on p. [22]. This constraint, like all the constraints on our list, was arrived at by attempting to set up the treatment of scientific revolutions in such a way that two desiderata would be achieved: (1) The system for evaluating scientific work and scientific standards should not be chauvinist — that is, it should not automatically accept or approve as reasonable only those explanations, theories, standards, and values which are acceptable to some one scientific community or which incorporate only the commitments of some particular conceputal scheme. (2) The system for evaluating science and scientific standards should be applicable to, and able to benefit from, detailed historical case studies of successful and unsuccessful science. It is of course not obvious that one can achieve these desiderata or that the six constraints I have proposed, if honored, would foster their realization. But the very fact that one can hope to aim at such a result suggests the value of explicit attention to (1) and (2). I suggest that they would serve us well as values and that we should attempt to realize them in our analyses of conceptual change. If one accepts this suggestion, then the application of the fifth constraint to rule out the quasi-intuitionist interpretation of scientific theories was entirely in order; if one seeks to realize desiderata (1) and (2) one should not accept an interpretation of scientific theories on which each theory "determines its own reality" unless one is forced to do so.

10. CONCLUSION

This paper yields two chief morals. The first is that numerous competing values underlie the alternative interpretations of theories and conceptual schemes. Since we understand very little about the relationships among these values, our analytical tools, and our interpretations of scientific theories, these relationships ought to become a prime target for philosophical investigation. The second is that if one values a "non-chauvinist" theory of conceptual change — specifically, one that does not suppose that we now generally possess the scientific truth where our ancestors did not — then one ought to develop analytical tools for studying the history of science that do not violate the six constraints set forth above.

Postscript, 1986. Since this article was first written, much of the philosophical writing touching on conceptual change has fit rather handily into the framework presented above. Although there has been somewhat less attention than I anticipated to the central role of values in the philosophical analysis of science (perhaps the most important exception is Laudan 1984), there has been quite considerable focus on the role of change-in-values as a contributing factor to change-of-theory and change-of-concept in science. (Unfortunately, far too much of this work has been "externalist" in spirit — cf. the development of the "strong programme" of Barnes, Bloor, Shapin *et al.* and the debate over it.) Of particular interest has been the criticism — and, indeed, the virtual abandonment — of holism with respect to meaning (cf., e.g., Glymour 1980 and Newton-Smith 1981, esp. Chap. VII) and the development of an alternative to holism with respect to the reference of theoretical terms (e.g., Cartwright 1983, Hacking 1983, Kitcher 1978 and 1985 and prior work on the so-called causal theory of reference.) Much, though by no means all, of this work may be viewed as an attempt to transcend or overcome the polarity between autonomism and historicism. One symptom of the retreat from autonomism (which I think is the dominant theme of the last five years) is the increasing proportion of work in the philosophy of science devoted to the philosophy of the particular sciences. These days, such attention to particular sciences is often highly "contextual" in character. That is, rather than applying *general* theories of confirmation, theory structure, or conceptual change to particular cases or sciences, many philosophers tend to treat their cases in ways that depend on the particular subject matter, background knowledge, the power of the available observational tools and theories, and so on. (So much for autonomism!) Yet much of this "contextualist" work is not strictly historicist either, for it seeks to evaluate evidence, theories, conceptual changes, and so on by use of philosophical and metascientific tools and modes of analysis that are not part of the armamentarium of particular scientific disciplines. (See Pitt 1985.) Philosophy of biology (which I know best) exemplifies this trend well. There is nowadays virtually no interest in showing that

— or specifically how — the structure of biological theories resembles that of the theories of physics. Indeed, many philosophers of biology are not persuaded of the structural similarity of theories in evolution, genetics, physiology, and systematics, to name but four fields. There is, therefore, considerably more concern than there used to be with understanding how such partially autonomous fields (and their allied theories and bodies of evidence) interact. (So much for holism!) Such diverse fields differ in their problems, their methodologies and methodological values, their theoretical and ontological commitments, yet they often make conflicting claims about the very same organisms. Accordingly, there is also considerable concern to work out an account of how the disagreements between such fields *ought* to be resolved. In the process, considerable effort has been expended in criticizing the conceptual foundations of particular disciplines and theories, such as Mendelian genetics, neo-Darwinian theories of natural selection, and phylogenetic systematics, and their mutual interactions. (So much for historicism!)

When philosophers of science deal with such context-specific issues, they tend to refine and differentiate their tools to suit the particular questions at hand. Thus, uniform analyses of the language of science, of the values of science, or of the ontological and theoretical commitments of science (or even, for that matter, of biology or physics) are less in evidence today than they were in 1980. (Perhaps the major exception here is found in the ongoing debate over scientific realism; cf., e.g., Leplin 1984, van Fraassen 1980, and the various critical analyses of the latter.) In some ways, therefore, such context specificity marks a retreat for philosophers: with it, the grand vision of a global theory of science and of scientific change seems to be receding further into the distance and into the past. But in other ways these developments are also salutory and encouraging; indeed that is the positive note on which I wish to close. It is a sign of maturity, marking a real advance, that philosophers of science are now interested in and able to address the particular problems faced by scientists (both living and from the distant past) in ways that take account of various changes in evidential relations, observational abilities, background knowledge, theoretical languages, mathematical tools, and so on. If we are to understand conceptual change in general, we will first have to understand much better than we now do the differences in the conceptual changes involved in the change from Maxwell to Einstein and that from Darwin to contemporary theorists of molecular evolution.

Virginia Polytechnic Institute
and State University

NOTES

* The following paper was completed in late 1979. The manuscript was delivered to the Pittsburgh Series in the Philosophy of Science in January, 1980; the editors of the Series offered assurances of reasonably prompt publication. The volume for which the paper was scheduled has now been postponed until (at least) 1988. Accordingly, to prevent further delay, I welcome the opportunity to publish the paper in the present volume. What follows is a lightly edited version of the 1980 text with a brief postscript commenting on subsequent developments.
1 I am grateful to Lorenz Krüger, Larry Laudan, Rachel Laudan, Jim Lennox, and Peter Machamer for criticisms of an earlier draft and especially to Catherine Elgin for an extensive critique that forced me to make a number of improvements in the paper. I am grateful to the National Endowment for the Humanities for a Summer Stipend in support of research.
2 This issue will be a major focus of discussion at a Conference on Testing Scientific Methodologies at the Center for the Study of Science in Society, Virginia Polytechnic Institute and State University, October, 1986.
3 Note that for a pure historicist, "are appropriate" should mean "are employed" or, perhaps, "may be self-consistently employed." Pure historicists do not claim to have the sort of independent authority that could establish that the standards in question *should* be employed in that context. In short, historicists describe a standard and the contexts in which it is used or consistently applicable, but they do not prescribe standards or independently evaluate the standards they describe.
4 Since most logical empiricists hold that acceptance of a theory is not justified unless, on the available evidence, it is highly probable that the theory is at least approximately true, it is often held that successful theories are only rarely genuinely abandoned. On such a view, superceded theories are typically subsumed in their successors.
5 This theme is developed in Hooker (1975). It is only fair to add that many logical empiricists sought to capture the standards they found in science rather than to derive, dictate, or legitimate scientific standards on the basis of considerations drawn from the theory knowledge.
6 Chapter XI of Kuhn (1970) is the *locus classicus* for support of the claim that science textbooks falsify the history of science.
7 An example developed in Judson (1979), esp. pp. 209–213, 215–216, and 608–613, concerns the shift-in-standards connected with the notion of biological specificity. The standard for the explanation of the specific action of a particular protein, for example, changed from determining its chemical formula to determining its sequence of

amino acids, to determining its three dimensional configuration (with charge distributions, etc.). At each stage, radical reevaluation of earlier work was required.

[8] Among others, Kuhn, Laudan, and Shapere have argued this point. References are supplied on page [4] above.

[9] Cf. esp. pp. 61—66. Cf. also Kuhn (1970), e.g., pp. 167 and 169, and Laudan (1977), pp. 74, 147 ff., 231 (note 5), 237 (note 18), and 238 (note 26).

[10] Ibid, esp. pp. 42—3, 58—9, 60. Compare Kuhn (1970), pp. 69—72, 99—100, 107, 130—135, 139, 148.

[11] I use the relatively neutral term "account" for this supposed explanation because of a dispute about its merits. At one pole are Kuhn and Laudan who treat the similarities among metals, respectively, as a "puzzle" or an abiding "empirical problem" solved by the phlogistic theory. (The puzzle counts as solved because the puzzling result was derived from premises consistent with the available data and congenial to the "paradigm" or "research tradition" involved. It does not matter whether the premises in question are true or even whether they have independent justification.) At the other pole are those who require explanations to have true or highly justified premises and who argue that the phlogistic account was never acceptable because there is no such substance as phlogiston and because the claim that phlogiston exists was never adequately justified. My own view of the merits of the phlogistic account is dim. As Alan Musgrave has pointed out to me in private conversation, many substances had in common a supposed excess of phlogiston, but only a few of these were metals. Why did the rest not behave as metals? Proponents of a position like that of Laudan and Kuhn maintain that this case offers an instance of loss-of-data, proponents of the second position are sceptical whether the data were ever properly encompassed within the phlogiston theory.

[12] Though the precise form the difficulty takes is disputed; is it that the supposed explanation was shown to be unsound or that the price (in anomalies, internal inconsistencies in the theory, unsolved conceptual problems, and so on) of accepting this successful "puzzle solution" became too high?

[13] Note that the issue is not whether quantitative chemistry is a clear-cut improvement on its predecessors; my text suggests how to argue that it was. Rather it is whether and how one can justify the shift-in-standards which in fact took place in this instance and the similar shifts which turn up ubiquitously in the case-study literature.

[14] Doppelt (1978), pp. 60—79, makes a strong case for the position that short-term losses need not, in the long run, be accepted into the body of science.

[15] The best general formalization of the theory of natural selection, that of Williams (1970, 1973), still does not integrate it with a theory of heredity.

[16] That is, not fixed, deductively closed bodies of propositions, but adjustable, frequently inconsistent sets of claims with a "program" or a priority order for making adjustments in the face of various sorts of difficulty.

[17] Note the extreme difference between asking "According to the explanatory ideals of systematists in England as of 1880, should Darwinism have been accepted?" — a question with a factual answer — and a question like the one in the text which is not susceptible of a purely factual answer.

[18] This argument for incommensurability depends on the implausible view that the total meaning of terms stems from the theory in which those terms are embedded. As

many critics have pointed out (e.g., recently, Leplin 1979), if only part of the meaning of a term is determined by the theory in which it is embedded, the argument breaks down.

[19] Doppelt (1978), has found a way of reading Kuhn on incommensurability that does not raise the difficulties I am discussing. But as Kuhn's discussion is usually read, it does in fact produce just these difficulties.

[20] In my (1985), esp. pp. 26—35, I suggest an approach to local comparison. As is clear from that text, I was strongly influenced by Kitcher (1978 and 1985).

[21] This does not mean that local comparison is *always* possible, nor that locally differing predictions need be practically testable. Thus even when local comparison is possible, it may not be feasible to execute the comparison.

[22] A local comparison of rival *versions* of competing theories, even when it decisively favors one *version* over the other, does not, of course, show which *theory* (if either) is correct or better supported. (This is part of the point of Lakatos's (1978) well-known argument against "instant rationality.") It remains possible that theories (as opposed to theory versions) can only be compared globally.

[23] I have presupposed throughout that we are dealing with science developed by human beings. The theoretical possibility that our science might be incommensurable with that of some alternative galactic culture (if we can make sense of such a possibility — cf. Davidson 1974) is irrelevant to the evaluation of the highest-level theories of human beings.

[24] Catherine Elgin's trenchant criticisms have greatly improved the next few paragraphs. I still have not resolved many of the difficulties that she has posed to me.

[25] Putnam 1978, pp. 28—29. The axioms in question here may be interpreted as including a group of "observation statements." The total axiom system then represents the accepted theory plus the accepted observation claims needed to establish initial conditions and provide evidential support. [Added in 1986: In an unpublished lecture, entitled "What's Wrong with the Model-Theoretic Argument Against Realism," Alan Musgrave argues covincingly that Putnam's procedure does not properly interpret the pre-theoretic or observation claims in question. Although Musgrave's results undermine many of Putnam's arguments for "internal realism," they in no way alter — indeed, they reinforce — what I claim above about the limitations of extensional languages plus formal semantics as analytical tools. The acceptability of an interpretation cannot be determined solely by formal means.]

[26] This claim may be read as a special instance of Quinean indeterminacy of translation.

[27] The problematic relation of a rational reconstruction to its original does not necessarily make the reconstruction useless. For example, examination of a force-free reconstruction of Newtonian mechanics might convince philosophers or scientists that even if the theory could be developed in a force-free manner, the resultant loss of empirical or heuristic power, the poor fit of the result with the "metaphysics" of the Newtonian tradition, etc., would render the resultant theory in important ways uninteresting. Such results can be quite important.

REFERENCES

Burian,, Richard. (1977). "More Than a Marriage of Convenience: On the Inextricability of History and Philosophy of Science." *Philosophy of Science* **44**: 1—42.
Burian, Richard. (1980). "Why Philosophers Should Not Despair of Understanding Scientific Discovery." In *Scientific Discovery, Logic, and Rationality.* Edited by Thomas Nickles. Dordrecht: D. Reidel. Pp. 317—336.
Burian, Richard. (1985). "On Conceptual Change in Biology: The Case of the Gene." In *Evolution at a Crossroads: The New Biology and the New Philosophy of Science.* Edited by David J. Depew and Bruce H. Weber. Cambridge, Mass.: MIT Press. Pp. 21—42.
Cartwright, Nancy. (1983). *How the Laws of Physics Lie.* Oxford: Clarendon Press.
Davidson, Donald. (1974). "On the Very Idea of a Conceptual Scheme." *Proceedings and Addresses of the American Philosophical Association* **47**: 5—20.
Doppelt, Gerald. (1978). "Kuhn's Epistemological Relativism: An Interpretation and Defense." *Inquiry* **21**: 33—86.
Feyerabend, Paul. (1975). *Against Method.* London: New Left Books.
Glymour, Clark. (1980). *Theory and Evidence.* Princeton: Princeton University Press.
Hacking, Ian. (1983). *Representing and Intervening.* Cambridge: Cambridge University Press.
Hooker, Cliff. (1975). "Philosophy and Meta-Philosophy of Science: Empiricism, Popperianism and Realism." *Synthese* **32**: 117—231.
Holton, Gerald. (1973). *Thematic Origins of Scientific Thought.* Cambridge, Mass: Harvard University Press.
Judson, Horace F. (1979). *The Eighth Day of Creation.* New York: Simon and Schuster.
Kitcher, Philip. (1978). "Theories, Theorists, and Theory Change." *Philosophical Review* **87**: 519—547.
Kitcher, Philip. (1985). "Genes." *British Journal for the Philosophy of Science* **33**: 337—359.
Kuhn, Thomas. (1970). *The Structure of Scientific Revolutions,* 2nd ed. Chicago: University of Chicago Press.
Kuhn, Thomas. (1977). *The Essential Tension: Selected Studies in Scientific Tradition and Change.* Chicago: University of Chicago Press.
Lakatos, Imre. (1978). "The Methodology of Scientific Research Programmes." In *Philosophical Papers,* vol. 1. Edited by John Worrall and Gregory Currie. Cambridge: Cambridge University Press. Pp. 8—101.
Laudan, Larry. (1977). *Progress and Its Problems: Towards a Theory of Scientific Growth.* Berkeley: University of California Press.
Laudan, Larry. (1979). "Historical Methodologics. An Overview and a Manifesto." In *Current Research in the Philosophy of Science.* Edited by Peter Asquith and Henry Kyburg. East Lansing, Mich: Philosophy of Science Assoication. Pp. 40—54.
Laudan, Larry. (1984). *Science and Values.* Berkeley: University of California Press.
Leplin, Jarrett. (1979). "Reference and Scientific Realism." *Studies in History and Philosophy of Science* **10**: 265—284.
Leplin, Jarrett (ed.). (1984). *Scientific Realism.* Berkeley: University of California Press.
Levin, Michael. (1979). "On Theory-Change and Meaning Change." *Philosophy of Science* **46**: 407—424.

CONCEPTUAL CHANGE

33

McMullin, Ernan. (1979). "The Ambiguity of 'Historicism'." In *Current Research in the Philosophy of Science*. Edited by Peter Asquith and Henry Kyburg. East Lansing, Mich: Philosophy of Science Association. Pp. 55—83.

Musgrave, Alan. (1983). "Facts and Values in Science Studies." In *Science Under Scrutiny: The Place of History and Philosophy of Science*. Edited by R. W. Home. Dordrecht: D. Reidel. Pp. 49—79.

Newton-Smith, W. H. (1981). *The Rationality of Science*. London: Routledge and Kegan Paul.

Nickles, Thomas. (1980). "Introductory Essay: Scientific Discovery and the Future of Philosophy of Science." In *Scientific Discovery, Logic, and Rationality*. Edited by Thomas Nickles. Dordrecht: D. Reidel. Pp. 1—59.

Pitt, Joseph C. (ed.). (1985). *Change and Progress in Modern Science*. Dordrecht: D. Reidel.

Putnam, Hilary. (1975). *Philosophical Papers*, vols. 1 and 2. Cambridge: Cambridge University Press.

Putnam, Hilary. (1978). *Meaning and the Moral Sciences*. London: Henley and Boston: Routledge and Kegan Paul.

Sellars, Wilfrid. (1963). *Science, Perception and Reality*. London: Routledge and Kegan Paul.

Sellars, Wilfrid. (1968). *Science and Metaphysics*. London: Routledge and Kegan Paul.

Sellars, Wilfrid. (1973). "Conceptual Change." In *Conceptual Change*. Edited by Glenn Pearce and Patrick Maynard. Dordrecht: D. Reidel. Pp. 77—93.

Shapere, Dudley. (1974a). "On the Relations Between Compositional and Evolutionary Theories." In *Studies in the Philosophy of Biology*. Edited by Francisco Ayala and Theodosius Dobzhansky. Berkeley and Los Angeles: University of California Press. Pp. 187—204.

Shapere, Dudley. (1974b). "Scientific Theories and Their Domains." In *The Structure of Scientific Theories*. Edited by Frederick Suppe. Urbana, Ill: University of Illinois Press. Pp. 518—564.

Shapere, Dudley. (1980). "The Character of Scientific Change." In *Scientific Discovery, Logic and Rationality*. Edited by Thomas Nickles. Dordrecht: D. Reidel. Pp. 61—101.

Stegmuller, Wolfgang. (1967). *The Structure and Dynamics of Theories*. New York: Springer Verlag.

Toulmin, Stephen. (1972). *Human Understanding*. Princeton: Princeton University Press.

Van Fraassen, Bas C. (1980). *The Scientific Image*. Oxford: The Clarendon Press.

Williams, Mary B. (1970). "Deducing the Consequences of Evolution." *Journal of Theoretical Biology* 29: 343—385.

Williams, Mary B. (1973). "Falsifiable Predictions of Evolutionary Theory." *Philosophy of Science* 40: 518—537.

KARL POPPER

THE MYTH OF THE FRAMEWORK*

"Those who believe this, and those who do not, have
no common ground of discussion, but in view of their
opinions they must of necessity scorn each other."

PLATO

I

One of the more disturbing features of intellectual life at the present
time is the way in which irrationalism is so widely advocated, and
irrationalist doctrines taken for granted. In my view, one of the main
components of modern irrationalism is relativism (the doctrine that
truth is relative to our intellectual background or framework: that it
may change from one framework to another), and, in particular, the
doctrine of the impossibility of mutual understanding between different
cultures, generations, or historical periods. In this paper I discuss the
problem of relativism. It is my claim that behind it lies what I call 'The
Myth of the Framework'. I explain and criticize this myth, and comment
also on arguments due to Quine, Kuhn, and Whorf which have been
used in its defence.

The proponents of relativism put before us standards of mutual
understanding which are unrealistically high; and when we fail to meet
those standards, they claim that understanding is impossible. Against
this, I argue that if common goodwill and a lot of effort are put into it,
then very far-reaching understanding is possible. Furthermore, the
effort is amply rewarded by what we learn in the process about our
own views, as well as about those we are setting out to understand.

This paper sets out to challenge relativism in its widest sense. It is
important to present such a challenge. For today, the increasing escala-
tion in the production of weapons has made survival almost identical
with understanding.

II

Although I am an admirer of tradition I am, at the same time, an almost

35

J. C. Pitt and M. Pera (eds.), Rational Changes in Science, 35—62.
© Sir Karl Popper 1976.

orthodox adherent of unorthodoxy: I hold that orthodoxy is the death of knowledge, since the growth of knowledge depends entirely on the existence of disagreement. Admittedly, disagreement *may* lead to strife, and even to violence; and this, I think, is very bad indeed, for I abhor violence. Yet disagreement may also lead to discussion, to argument — to mutual criticism — and this, I think, is of paramount importance. I suggest that the greatest step towards a better and more peaceful world was taken when the war of swords began to be supported, and sometimes even to be replaced, by a war of words. This is why my topic is of practical significance.

But let me first explain what my topic is, and what I mean by my title, 'The Myth of the Framework'. I will discuss, and argue against, a myth — a false story that is widely accepted, especially in Germany. From there it invaded America where it became almost all-pervasive. So I fear that the majority of my present readers may believe in it, either consciously or unconsciously. The myth of the framework can be stated in one sentence, as follows:

A rational and fruitful discussion is impossible unless the participants share a common framework of basic assumptions or, at least, unless they have agreed on such a framework for the purpose of the discussion.

This is the myth I am going to criticize.

As I have formulated it, the myth sounds like a sober statement, or like a sensible warning to which we ought to pay attention in order to further rational discussion. Some people even think that it is a logical principle, or based on a logical principle. On the contrary, I think that it is not only a false statement but also a vicious statement which, if widely believed, must undermine the unity of mankind, and must greatly increase the likelihood of violence and of war. This is the main reason why I want to combat it, and to refute it.

Let me say at once that the myth contains a kernel of truth. Although I contend that it is a vast exaggeration to say that a fruitful discussion is *impossible* unless the participants share a common framework, I am very ready to admit that a discussion among participants who do not share a common framework may be *difficult*. A discussion will also be difficult if the frameworks have little in common, and it will be the easier the greater the overlap between the frameworks. Indeed, if the participants agree on all points, it will often turn out to be the easiest and smoothest discussion possible — though it is likely to be a little boring.

But what about fruitfulness? In the formulation I gave of the myth, it is a *fruitful* discussion which is declared impossible. Against this I shall defend the thesis that a discussion between people who share many views is unlikely to be fruitful, even though they may regard it as pleasant and highly satisfactory, while a discussion between vastly different frameworks can be extremely fruitful even though it will usually be difficult and *perhaps* not quite so pleasant (though we may learn to enjoy it).

I think that we may say of a discussion that it was the more fruitful the more its participants learned from it. And this means: the more interesting questions and difficult questions they were asked; the more new answers they were induced to think of; the more they were shaken in their opinions; and the more they could see things differently after the discussion; in short, the more their intellectual horizon was extended.

Fruitfulness in this sense will almost always depend on the original gap between the opinions of the participants in the discussion. The greater the gap, the more fruitful *can* the discussion be — always provided of course that such a discussion is not altogether *impossible*, as the myth of the framework asserts.

III

But is it impossible? Let us take an extreme case. Herodotus tells a very interesting though somewhat gruesome story of the Persian King Darius the First who wanted to teach a lesson to the Greek residents in his country, whose custom it was to burn their dead. He "summoned", we read in Herodotus, "the Greeks living in his land, and asked them for what payment they would consent to eat up their fathers when they died. They answered that nothing on earth would induce them to do so. Then Darius summoned the ... Callatians, who do eat their fathers, and asked them in the presence of the Greeks, who had the help of an interpreter, for what payment they would consent to burn the bodies of their fathers when they died. And they cried out aloud and implored him not to mention such an abomination." [1]

Darius, I suspect, wanted to demonstrate the truth of the myth of the framework. Indeed, we are given to understand that a discussion between the two parties would have been impossible even with the help of the interpreter. It was an extreme case of a '*confrontation*' — to use a word much in vogue with believers in the truth of the myth, and a word they like to use when they wish to draw our attention to the fact that a confrontation rarely results in a fruitful discussion.

But assuming that this confrontation staged by King Darius did take place, was it really fruitless? I deny it. There can be little doubt that both parties were deeply shaken by the experience. I myself find the idea of cannibalism just as revolting as did the Greeks at the court of King Darius, and I suppose my readers will feel the same. But these feelings should make us all the more perceptive and the more appreciative of the admirable lesson which Herodotus wishes to draw from the story. Alluding to Pindar's distinction between nature and convention,[2] Herodotus suggests that we should look with tolerance and even with respect upon customs or conventional laws that differ from our own conventions. If this particular confrontation ever took place, some of the participants may well have reacted to it in the enlightened way in which Herodotus wishes us to react to his story.

This shows that there is a possibility of a fruitful confrontation, even without a discussion, of people deeply committed to different frameworks. Of course, *we must not expect too much*: we must *not* expect that a confrontation, or even a prolonged discussion, will end with the participants reaching *agreement*.

But is an agreement always desirable? Let us assume that there is a discussion and that the issue at stake is the truth or falsity of some theory or hypothesis. We — that is, the rational witnesses or judges of the discussion — would of course like the discussion to end with all parties agreeing that the theory is true if in fact it is true, or that the theory is false if in fact it is false: we should like the discussion to reach, if possible, a true verdict. But we should dislike the idea that agreement was reached on the truth of the theory if the theory was in fact false; and even if it was true, we prefer that no agreement is reached on its truth if the arguments supporting the theory were far too weak to bear out the conclusion. In such a case we prefer that no agreement is reached. And in such a case we should say that the discussion was fruitful when the clash of opinion led the participants to produce new and interesting arguments, even though these arguments were inconclusive. For conclusive arguments are very rare in all but the most trivial issues, even though arguments against a theory may sometimes be pretty strong.

Looking back at Herodotus's story of a confrontation, we can now see that even in this extreme case where no agreement was in sight the confrontation may have been useful and that, given time and patience

— which Herodotus seems to have had at his disposal — it did bear fruit, at least in Herodotus's own mind.

IV

Now I wish to suggest that, in a way, we ourselves and our attitudes are the results of confrontations and of inconclusive discussions of this kind.

What I mean can be summed up by the thesis that our Western civilization is the result of the clash, or the confrontation, of different cultures, and therefore of the confrontation of frameworks.

It is widely admitted that our civilization — which at its best may be described, somewhat eulogistically, as a rationalist civilization — is very largely the result of Greco-Roman civilization. It acquired many of its features, such as the alphabet, and Christianity, not only through the clashes between the Romans and the Greeks, but also through its clashes with the Jewish, the Phoenician, and other Middle Eastern civilizations, and also through clashes due to Germanic and Islamic invasions.

But what of the original Greek miracle — the rise of Greek poetry, art, philosophy, and science; the real origin of Western rationalism? I have for many years asserted that the Greek miracle, *insofar as it can be explained*, was also largely due to culture clash. It seems to me that this is indeed one of the lessons which Herodotus wants to teach us in his *History*.

Let us look for a moment at the origin of Greek philosophy. It all began in the Greek colonies in Asia Minor, in Southern Italy, and in Sicily; places, that is, where, in the East, the Greek colonists were confronted with the great oriental civilizations, and clashed with them, or where, in the West, they met Sicilians, Carthaginians, and Italians such as the Tuscans. The impact of culture clash on Greek philosophy is very obvious from the earliest reports on Thales. It is unmistakable in Heraclitus. But the way in which it leads men to think critically comes out most forcefully in Xenophanes, the wandering bard. Although I have quoted some of his verses on other occasions, I will do so again, because they illustrate my point so beautifully.[3]

The Ethiops say that gods are flat-nosed and black
While the Thracians say that theirs have blue eyes and red hair.

Yet if cattle or horses or lions had hands and could draw
And could sculpture like men, then the horses would draw their gods
Like horses, and cattle like cattle, and each would then shape
Bodies of gods in the likeness, each kind, of its own.

The gods did not reveal, from the beginning,
All things to us; but in the course of time,
Through seeking we may learn, and know things better. . . .

These things are, we conjecture, like the truth.
But as for certain truth, no man has known it,
Nor will he know it; neither of the gods,
Nor yet of all the things of which I speak.
And even if by chance he were to utter
The final truth, he would himself not know it:
For all is but a woven web of guesses.

Although Burnet and others have denied it, I think that Parmenides, perhaps the greatest of these early thinkers, stood under Xenophanes' influence.[4] He takes up Xenophanes' distinction between the one final truth which is not subject to human convention, and the guesses or opinions, and the conventions, of the mortals. There are always many conflicting opinions and conventions concerning any one problem or subject matter (such as the gods), which shows that they are not all true, for if they conflict then, at best, only one of them can be true.[5] Thus it appears that Parmenides (a contemporary of Pindar to whom Plato attributes the distinction between nature and convention) was the first to distinguish clearly between truth or reality on the one hand, and convention or conventional opinion — hearsay, plausible myth — on the other; a lesson which, we may say, he derived from Xenophanes and from culture clash. It led him to one of the boldest theories ever conceived.

The role played by culture clash in the rise of Greek science — mathematics and astronomy — is well known, and one can even specify the way in which the various clashes bore fruit. And our ideas of freedom, of democracy, of toleration, and also the ideas of knowledge, of science, of rationality, can all be traced back to these beginnings.

Of all these ideas the idea of rationality seems to me the most fundamental.

So far as we know from the sources, the invention of rational or critical discussion seems to be contemporaneous with some of these clashes, and discussion became traditional with the rise of the earliest Ionian democracies.

V

In its application to the problem of understanding our world, and thus to the rise of science, rationality has two components which are of about equal importance.

The first is poetic inventiveness, that is, storytelling or mythmaking: the invention of stories which explain the world. These are, to begin with, often or perhaps always polytheistic. Men feel that they are in the hands of unknown powers, and they try to understand and to explain the world, and human life and death, by inventing stories or myths about these powers.

This first component, which may be perhaps as old as human language itself, is all-important and seems universal: all tribes, all peoples, have such explanatory stories, often in the form of fairy tales. It seems that the invention of explanations and explanatory stories is one of the basic functions of the human language.

The second component is of comparatively recent date. It seems to be specifically Greek and to have arisen after the establishment of writing in Greece. It arose, it seems, with Anaximander, the second Ionian philosopher. It is the invention of criticism, of the critical discussion of the various explanatory myths, with the aim of consciously improving upon them.

The main Greek example of explanatory mythmaking on an elaborate scale is, of course, Hesiod's *Theogony*. This is a wild story of the origin, the deeds, and the misdeeds, of the Greek gods. One would hardly feel inclined to look to the *Theogony* to provide a suggestion which can be used in the development of a scientific explanation of the world. Yet I have proposed the historical conjecture that a passage in Hesiod's *Theogony*[6] which was foreshadowed by another in Homer's *Iliad*[7] was so used by Anaximander, the first critical cosmologist.

I will explain my conjecture. According to tradition Thales, the teacher and kinsman of Anaximander, and the founder of the Ionian school of cosmologists, taught that 'the earth is supported by water on which it rides like a ship'. Anaximander, the pupil, kinsman, and successor of Thales, turned away from this somewhat naive myth (intended by Thales to explain earthquakes). Anaximander's new departure was of a truly revolutionary character, for he taught, we are told, the following: "There is no thing at all that is holding up the earth. Instead, the earth remains stationary owing to the fact that it is equally far away from all other things. Its shape is like that of a drum. We walk on one of its flat surfaces while the other is on the opposite side."

This bold idea made possible the ideas of Aristarchus and Copernicus, and it even contains an anticipation of Newton's forces. How did it arise? I have proposed the conjecture[8] that it arose out of a purely logical criticism of Thales' myth. The criticism is simple: if we solve the problem of explaining the position and stability of the earth in the universe by saying that it is supported by the ocean, like a ship that is supported by water, are we not then bound, the critic asks, to raise a new problem, that of explaining the position and the stability of the ocean? But this would mean finding some support for the ocean, and then some further support for this support. Obviously, this leads to an infinite regress. How can we avoid it?

In looking round for a way out of this frightful impasse which, it appeared, no alternative explanation was able to avoid, Anaximander remembered, I conjecture, a passage in which Hesiod develops an idea from the *Iliad* where we are told that Tartarus is exactly as far beneath the earth as Uranus, or heaven, is above it.

The passage reads: "For nine days and nights will a brazen anvil fall from the heavens, and on the tenth it will reach the earth. And for nine days and nights will a brazen anvil fall from the earth, and on the tenth it will reach Tartarus."[9] This passage may have suggested to Anaximander that we can draw a diagram of the world, with the earth in the middle, and the vault of the heavens like a hemisphere above it. Symmetry then suggests that we interpret Tartarus as being the lower half of the vault. In this way we arrive at Anaximander's construction as it is transmitted to us; a construction that breaks through the deadlock of the infinite regress.

There is I think a need for such a conjectural explanation of the tremendous step that carried Anaximander beyond his teacher Thales. My conjecture, it seems to me, makes the step more understandable and, at the same time, even more impressive; for it is now seen as a rational solution of a very difficult problem — the problem of the support and the stability of the earth.

Yet Anaximander's criticism of Thales and his critical construction of a new myth would have led to nothing had these not been followed up. How can we explain the fact that they *were* followed up? Why was a new myth offered in each generation after Thales? I have tried to explain this by the further conjecture that Thales and Anaximander together founded a new school tradition — *the critical tradition.*

My attempt to explain the phenomenon of Greek rationalism and of the Greek critical tradition by a school tradition is again, of course, completely conjectural. In fact, it is itself a kind of myth. Yet it does explain a unique phenomenon — the Ionian school. This school, for at least four or five generations, produced in each new generation an ingenious revision of the teachings of the preceding generation. In the end it established what we may call the scientific tradition: a tradition of criticism which survived for at least five hundred years, and which survived some serious onslaughts before it succumbed.

The critical tradition is constituted by the adoption of the method of criticizing a received story or explanation and then proceeding to a new, improved, imaginative story which in turn is submitted to criticism. This method, I assert, is the method of science. It seems to have been invented only once in human history. It died in the West when the schools in Athens were suppressed by a victorious and intolerant Christianity, though it lingered on in the East. It was mourned during the Middle Ages. And it was not so much reinvented as reimported in the Renaissance, together with the rediscovery of Greek philosophy and Greek science.

The uniqueness of this second component — the method of criticial discussion — will be realized if we consider the old-established function of schools, especially of religious and semireligious schools. Their function is, and has always been, the preservation of the purity of the teaching of the founder of the school. Accordingly, changes in doctrine are rare and are often due to mistakes or misunderstandings. When they are consciously made they are as a rule made surreptitiously; for otherwise changes lead to splits, to schisms.

But here, in the Ionian school, we find a school tradition which carefully preserved the teaching of each of its masters while deviating from it afresh in each new generation.

My conjectural explanation of this unique phenomenon is that Thales, the founder, encouraged Anaximander, his kinsman, pupil, and later his successor, to see whether he could produce a better explanation of the support of the earth than he himself had been able to offer.

However this may have been, the invention of the critical method could hardly have happened without the impact of culture clash. It had the most tremendous consequences. Within four or five generations the Greeks discovered that the earth, the moon, and the sun, were spheres; that the moon moved round the earth, while always 'wistfully' looking at the sun; and that this could be explained by the assumption that she borrowed her light from the sun.[10] A little later they conjectured that the earth rotated, and that the earth moved round the sun. But these later hypotheses, due to the Platonic school and especially to Aristarchus, were soon forgotten.

These cosmological or astronomical findings became the basis of all future science. Human science started from a bold and hopeful attempt to understand critically the world we live in. This ancient dream found fulfillment in Newton. We can say that only since Newton has humanity become fully conscious — conscious of its position in the universe.

All this, it can be shown, is the result of applying the method of critical discussion to mythmaking — to our attempts to understand and to explain our world.

VI

If we look back on this development, then we can understand better why we must not expect any critical discussion of a serious issue, any 'confrontation', to yield quick and final results. Truth is hard to come by. It needs both ingenuity in criticizing old theories, and ingenuity in the imaginative invention of new theories. This is so not only in the sciences, but in all fields.

Serious critical discussions are always difficult. Nonrational human elements always enter. Many participants in a rational, that is, a critical, discussion find it particularly difficult that they have to unlearn what everybody is taught in a debating society, for they have to learn that victory in a debate is nothing, while even the slightest clarification of one's problem, even the smallest contribution made towards a clearer understanding of one's own position or that of one's opponent, is a great success. A discussion which you win but which fails to help you to change or to clarify your mind at least a little should be regarded by you as a sheer loss. For this very reason no change in one's position should be made surreptitiously, but it should always be stressed, and its consequences explored.

Rational discussion in this sense is rare. But it is an important ideal, and we may learn to enjoy it. It does not aim at conversion, and it is modest in its expectations: it is enough, more than enough, if we feel that we can see things in a new light, or that we have got even a little nearer to the truth.

VII

But let me now return to the myth of the framework. There are many tendencies which may contribute to the fact that this myth is often taken for an almost self-evident truth.

One of these tendencies I have already mentioned. It results from an overoptimistic expectation concerning the outcome of a discussion; the expectation that every fruitful discussion should lead to decisive and deserved intellectual victory of the truth, represented by one part, over falsity, represented by the other. When it is found that this is not what a discussion usually achieves, disappointment turns an overoptimistic expectation into a general pessimism concerning the value of discussions.

A second tendency which deserves careful scrutiny is connected with historical or cultural relativism, a view whose beginnings may perhaps be discerned in Herodotus, the father of history.

Herodotus seems to have been one of those somewhat uncommon people whose mind was broadened by travel. At first he was no doubt shocked by the many strange customs and institutions which he encountered in the East. But he learned to respect them, and to look on some of them critically, on others as the results of historical accidents: he learned to be tolerant, and he even acquired the ability to see the customs and institutions of his own country through the eyes of his barbarian hosts.

This is a healthy state of affairs. But it may lead to relativism, this is, to the view that there is no absolute or objective truth, but rather one truth for the Greeks, and another for the Egyptians, and still another for the Syrians, and so on.

I do not think that Herodotus fell into this trap. But many have done so since — perhaps inspired by an admirable feeling of tolerance which they have combined with very dubious logic.

There is one version of the idea of cultural relativism which is obviously correct. In England, Australia, and New Zealand we drive on the left-hand side of the road, while in America and in most other countries we drive on the right-hand side. What is needed is *some* such rule of the road, but which of the two — the right or the left — is obviously arbitrary and conventional. There are many similar rules of greater or lesser importance which are purely conventional or customary. Among these are the different rules for pronouncing and spelling the English language in America and in England. Even two quite different vocabularies may be related in a conventional way closely resembling the two different rules of the road, provided the grammatical structures of the two languages are very similar. We may regard such vocabularies, or such rules, as differing in a purely conventional way: there is really nothing to choose between them — nothing of importance.

As long as we consider only conventional rules and customs such as these, there is no chance for the myth of the framework to be taken seriously; for a discussion between an American and an Englishman about the rule of the road is likely to lead to an agreement. Both are likely to regret the fact that their rules do not coincide. Both will agree that in principle there is nothing to choose between the two rules, and that it would be unreasonable to expect the United States to adopt the left-hand rule in order to achieve conformity with Britain; and both are likely to agree that Britain cannot at present make a change which may be desirable but which would be extremely costly. After agreement has thus been reached on all points, both participants are likely to part with the feeling that they have not learned anything from the discussion.

The situation changes when we consider other institutions, laws, or customs — those for example which are connected with the administration of justice. Different laws and customs in this field may make all the difference for those living under them. Some customs can be very cruel, while others provide for mutual help and the relief of suffering. Some countries and their laws respect freedom while others do so less, or not at all.

It is my opinion that a critical discussion of these important matters is not only possible, but most urgently needed. It is often made difficult by propaganda and by a neglect of factual information. But these difficulties are not insuperable. Thus it is possible to combat propaganda by information, and information, if available, is not always ignored; though admittedly it often is ignored.

In spite of all this there are some people who uphold the myth that frameworks of laws and customs cannot be rationally discussed. They assert that morality is identical with legality or custom or usage, and that it is therefore impossible to judge, or discuss, whether one system of customs is morally better than another, since the existing system of laws and customs is the only possible standard of morality.

This view has been stated by Hegel with the help of the formulae: 'What is real is reasonable' and 'What is reasonable is real'. Here 'what is' or 'what is real' means the world, including its man-made laws and customs. That these are man-made is denied by Hegel who asserts that the World Spirit or Reason made them, and that those who seem to have made them — the great men, the makers of history — are merely the executors of reason, their passions being the most sensitive instrument of reason; they are the detectors of the Spirit of their Time, and ultimately of the Absolute Spirit, that is of God Himself.

This is just one of those many cases in which philosophers use God for their own private purposes; that is, as a prop for some of their tottering arguments.

Hegel was both a relativist and an absolutist: as always, he had it at least both ways, and if two ways were not enough, he had it in three ways. And he was the first of a long chain of post-Kantian, that is, postcritical or postrationalist philosophers — mainly German philosophers — who upheld the myth of the framework.

According to Hegel, truth itself was both relative and absolute. It was relative to each historical and cultural framework: there could thus be no rational discussion between the frameworks since each of them had a different standard of truth. But his doctrine that all truth was relative to the various frameworks was absolutely true, since it was part of Hegel's own relativistic philosophy.

VIII

Hegel's claim to have discovered absolute truth does not now appear to attract many people. But his doctrine of relative truth and his myth of the framework still attracts them. What makes it so attractive is that they confuse relativism with the true insight that all men are fallible. This doctrine of fallibility has played an important role in the history of philosophy from its earliest days on — from Xenophanes and Socrates to Charles Sanders Peirce — and I think that it is of the utmost importance. But I do *not* think that it can be used to support relativism with respect to truth.

Of course, the doctrine of human fallibility can be validly used to argue against that kind of philosophical absolutism which claims to *possess* the absolute truth, or at least a criterion of absolute truth, such as the Cartesian criterion of clarity and distinctness, or some other intuitive criterion. But there exists a very different doctrine of absolute truth, in fact a fallibilist doctrine, which asserts that mistakes we make can be absolute mistakes, in the sense that our theories can be absolutely false, that they can fall short of the truth. Thus the notion of truth, and that of falling short of the truth, can represent absolute standards for the fallibilist. These notions are a great help in critical discussions.

This theory of absolute or objective truth has been revived by Alfred Tarski who also proved that there can be no universal criterion of truth. There is no clash whatever between Tarski's theory of absolute or objective truth and the doctrine of fallibility.[11]

But is not Tarski's notion of truth a relative notion? Is it not relative to the language to which the statement belongs whose truth is being discussed?

The answer to this question is 'no'. Tarski's theory says that a statement of some language, say English, is true if and only if it corresponds to the facts; and Tarski's theory implies that whenever there is another language, say French, in which we can describe the same fact, then the French statement which describes this fact will be true if and only if the corresponding English statement is true. Thus it is impossible, according to Tarski's theory, that of two statements which are translations of each other, one can be true and the other false. Truth, according to Tarski's theory, is therefore *not* dependent on language, or relative to language. Reference to the language is made only because of the unlikely but trivial possibility that the same sounds or symbols may occur in two different languages and may then perhaps describe two totally different facts.

However, it may easily happen that a statement of one language is untranslatable into another, or in other words that a fact, or a state of affairs, which can be described in one language cannot be described in another.

Anybody who can speak more than one language knows, of course, that perfect translations from one language into another are very rare, if they exist at all. But this difficulty, well-known to all translators, should be clearly distinguished from the situation here discussed — that is, the impossibility of describing in one language a state of affairs which can be described in some other language. The ordinary and well-known difficulty consists of something quite different, namely this. A crisp, simple, and easily understandable statement in French or English may need a highly complex and awkward rendering in, say, German, and a rendering which is even difficult to understand in German. In other words, the ordinary difficulty known to every translator is that an aesthetically adequate translation may be impossible, not that *any* translation of the statement in question is impossible. (I am speaking here of a factual statement, not of a poem or an aphorism or bon mot, or of a statement which is subtly ironical or which expresses a sentiment of the speaker.)

There can be no doubt, however, that a more radical impossibility may arise; for example, we can construct artificial languages which contain only one-termed predicates, so that we can say in these languages 'Paul is tall' and 'Peter is short', but not 'Paul is taller than Peter'.

More interesting than such artifical languages are some living languages. Here we can learn much from Benjamin Lee Whorf.[12] Whorf was perhaps the first to draw attention to the significance of certain tenses of the Hopi language. These tenses are experienced by a Hopi speaker as describing some part of the state of affairs which he tries to describe in his statement. They cannot be adequately rendered into English, for we can explain them only in a roundabout way, by referring to certain expectations of the speaker rather than aspects of the objective states of affairs.

Whorf gives the following example. There are two tenses in Hopi which might inadequately be rendered in English by the two statements

'Fred began chopping wood', and
'Fred began to chop wood'.

The first would be used by the Hopi speaker if he expects Fred to *go on* chopping for some time. If the speaker does *not* expect Fred to go on chopping, then he will *not* say, in Hopi, 'Fred began chopping'; he will use that other tense rendered by 'Fred began to chop'. But the real point is that the Hopi speaker does not wish by the use of his tenses merely to express his different expectations. He rather wishes to describe two different states of affairs — two different objective situations, two different states of the objective world. The one tense may be said to describe the beginning of a continuing *state* or of a somewhat repetitive *process*, while the other describes the beginning of an *event* of short duration. Thus the Hopi speaker may try to translate Hopi into English by saying: 'Fred began sleeping', in contradistinction to 'Fred began to sleep', because sleeping is a process rather than an event.

All this is very much simplified: a full restatement of Whorf's description of the complex linguistic situation could easily take up a whole paper. The main consequence for my topic which seems to emerge from the situations described by Whorf and more recently discussed by Quine is this. Although there cannot be any linguistic relativity concerning the *truth* of any statement, there is the possibility that a statement may be untranslatable into some other language. For the two languages may have built into their very grammar two different views of the stuff the world is made of, or of the world's basic structural characteristics. In the terminology of Quine this may be called the 'ontological relativity' of language.[13]

The possibility that some statements are untranslatable is, I assert, about the most radical consequence we can draw from what Quine calls 'ontological relativity'. Yet in actual fact most human languages seem to be intertranslatable. We may say that they are mostly *badly* inter-translatable, mainly because of ontological relativity, although of course for other reasons too. For example, appeals to our sense of humour, or comparisons with a well-known local or historical event which has become typical may be completely untranslatable.

IX

It is obvious that this situation must make rational discussion very difficult if the participants are brought up in different parts of the world, and speak different languages. But I have found that these difficulties can often be surmounted. I have had students in the London School of Economics from various parts of Africa, the Middle East, India, Southeast Asia, China, and Japan, and I have found that the difficulties could usually be conquered with a little patience on both sides. Whenever there was a major obstacle to overcome, it was as a rule the result of indoctrination with Western ideas. Dogmatic, uncritical teaching in bad Westernized schools and universities, and especially training in Western verbosity and in Western ideologies were, in my experience, much graver obstacles to rational discussion than any cultural or linguistic gap.

My experiences suggested to me that culture clash may lose some of its value if one of the clashing cultures regards itself as universally superior, and even more so if it is so regarded by the other: this destroys the major value of culture clash, for the greatest value of culture clash lies in the fact that it can evoke a critical attitude. More especially, if one of the parties becomes convinced of his inferiority, then the critical attitude of learning from the other will be replaced by a kind of blind acceptance, a blind leap into a new magic circle, or a conversion, as it is so often described by fideists and existentialists.

I believe that ontological relativity, though an obstacle to easy communication, can prove of immense value in all the more important cases of culture clash if it can be overcome slowly. For it means that the partners in the clash may liberate themselves from prejudices of which they are unconscious — from taking theories unconsciously for granted which, for example, may be embedded in the logical structure of their language. Such a liberation may be the result of *criticism* stimulated by culture clash.

What happens in such cases? We compare and contrast the new language with our own, or with some others we know well. In the comparative study of these languages we use, as a rule, our own language as a metalanguage — that is, as the language in which we speak about, and compare, the other languages which are the objects under investigation, including our own language. The languages under investigation are the object languages. In carrying out the investigation, we are forced to look upon our own language — say English — in a critical way, as a set of rules and usages which may be somewhat narrow since they are unable completely to capture, or to describe, the kinds of entities which the other languages assume to exist. But this description of the limitations of English as an object language is carried out in English as a metalanguage. Thus we are forced, by this comparative study, to transcend precisely those limitations which we are studying. And the interesting point is that we succeed in this. The means of transcending our language is *criticism*.

Whorf himself, and some of his followers, have suggested that we live in a kind of intellectual prison, a prison formed by the structural rules of our language. I am prepared to accept this metaphor, though I have to add to it that it is an odd prison insofar as we are normally unaware of it. We become aware of it through culture clash. But then, this very awareness allows us to break out of the prison if we wish to: we can transcend our prison by studying the new language and comparing it with our own.

The result will be a new prison. But it will be a much larger and wider prison; and again, we will not suffer from it; or rather, whenever we do, we are free to examine it critically, and thus to break out again, into a still wider prison.

The prisons are the frameworks. And those who do not like prisons will be opposed to the myth of the framework. They will welcome a discussion with a partner who comes from another world, from another framework, for it gives them an opportunity to discover their so far unfelt chains, to break them, and thus to transcend themselves. This breaking out of one's prison is, of course, not a matter of routine:[13a] it can only be the result of a critical effort — of a creative effort.

X

In the remainder of this paper I will try to apply this brief analysis to some problems which have arisen in a field in which I am greatly interested — the philosophy of science.

It is now fifty year since I arrived at a view very similar to the myth of the framework; and I not only arrived at it but at once went beyond it. It was during the great and heated discussions after the First World War that I found out how difficult it was to get anywhere with people living in a closed framework; I mean people like the Marxists, the Freudians, and the Adlerians. None of them could ever be shaken in his adopted view of the world. Every argument against their framework was by them so interpreted as to fit into it; and if this turned out to be difficult, then it was always possible to psychoanalyse or socioanalyse the arguer: criticism of Marxian ideas was due to class prejudice, criticism of Freudian ideas was due to repression, and criticism of Adlerian ideas was due to the urge to prove your superiority, an urge which was due to an attempt to compensate for a feeling of inferiority.

I found the stereotyped pattern of these attitudes depressing and repelling, the more so as I could find nothing of the kind in the debates of the physicists about Einstein's General Theory, although it too was hotly debated at the time.

The lesson I derived from these experiences was this. Theories are important and indispensable because without them we could not orientate ourselves in the world — we could not live. Even our observations are interpreted with their help. The Marxist literally sees class struggle everywhere; thus he believes that only those who deliberately shut their eyes can fail to see it. The Freudian sees everywhere repression and sublimation; the Adlerian sees how feelings of inferiority express themselves in every action and every utterance, whether it is an utterance of inferiority or superiority.

This shows that our need for theories is immense, and so is the power of theories. Thus it is all the more important to guard against becoming addicted to any particular theory: we must not let ourselves be caught in a mental prison. I did not know of the theory of culture clash at the time, but I certainly made use of my clashes with the addicts of the various frameworks in order to impress upon my mind the ideal of liberating oneself from the intellectual prison of a theory in which one might get stuck unconsciously, at any moment of one's life.

It is only too obvious that this ideal of self-liberation, of breaking out of one's prison of the moment, might in its turn become part of a framework or a prison — or in other words, that we can never be absolutely free. But we can widen our prison, and at least we can leave behind the narrowness of one who is addicted to his fetters.

Thus our view of the world is at any moment necessarily theory impregnated. But this does not prevent us from progressing to better theories. How do we do it? The essential step is the linguistic formulation of our beliefs. This objectivizes them; and this makes it possible for them to become targets of criticism. Thus our beliefs are replaced by competing theories, by competing conjectures. And through the critical discussion of these theories we can progress.

In this way we must demand of any better theory, that is, of any theory which may be regarded as progressing beyond some less good theory, that is can be compared with the latter. In other words, that the two theories are *not* 'incommensurable', to use a now fashionable term, introduced in this context by Thomas Kuhn.

(Note that two logically incompatible theories will be, in general, 'commensurable'. *Incommensurability* is intended to be much more radical than *incompatibility*: while incompatibility is a logical relation and thus appeals to one logical framework, incommensurability suggests the non-existence of a common logical framework.)

For example, Ptolemy's astronomy is far from incommensurable with that of Aristarchus and Copernicus. No doubt, the Copernican system allows us to see the world in a totally different way; no doubt there is, psychologically, a *Gestalt* switch, as Kuhn calls it. This is psychologically very important. But we *can* compare the two systems logically. In fact, it was one of Copernicus's main arguments that all astronomical observations which can be fitted into a geocentric system can, by a simple translation method, always be fitted into a heliocentric one. There is no doubt all the difference in the world between these two views of the universe, and the magnitude of the gulf between the two views may well make us tremble. But there is no difficulty in comparing them. For example, we may point out the colossal velocities which the rotating sphere of the fixed stars must give to the stars which are near to its equator, while the rotation of the earth, which in Copernicus's system replaces that of the fixed stars, involves very much smaller velocities. This, together with some practical acquaintance with centrifugal forces, may well have served as an important point of comparison for those who had to choose between the two systems.

I assert that this kind of comparison between systems is always possible. Theories which offer solutions of the same or closely related problems are as a rule comparable, I assert, and discussions between them are always possible and fruitful; and not only are they possible, but they actually take place.

XI

Some people do not think that these assertions are correct, and this results in a view of science and its history very different from mine. Let me briefly outline such a view of science.

The proponents[14] of such a view can observe that scientists are, normally, engaged in close cooperation and discussion; and the proponents argue that this situation is made possible by the fact that scientists normally operate within a common framework to which they have committed themselves. (Frameworks of this kind seem to me to be closely related to what Karl Mannheim used to call 'Total Ideologies'.[15]) The periods during which scientists remain committed to a framework are regarded as typical: they are periods of 'normal science', and scientists who work in this way are regarded as 'normal scientists'.

Science in this sense is then contrasted with science in a period of crisis or revolution. These are periods in which the theoretical framework begins to crack, and in the end breaks. It is then replaced by a new one. The transition from an old framework to a new one is regarded as a process which must be studied not from a logical point of view (for it is, essentially, not wholly, or even mainly, rational) but from a psychological and sociological point of view. There is, perhaps, something like 'progress' in the transition to a new theoretical framework. But this is not a progress which consists of getting nearer to the truth, and the transition is not guided by a rational discussion of the relative merits of the competing theories. *It cannot be so guided since a genuinely rational discussion is thought to be impossible without an established framework.* Without a framework it is not even thought to be possible to agree what constitutes a point of 'merit' in a theory. (Some protagonists of this view even think that we can speak of truth only relative to a framework.) Rational discussion is thus impossible if it is the framework which is being challenged. And this is why the two frameworks — the old and the new — have sometimes been described as *incommensurable*.

An additional reason why frameworks are sometimes said to be incommensurable seems to be this. A framework can be thought of as consisting not only of a 'dominant theory', but also as being, in part, a psychological and sociological entity. It consists of a dominant theory *together* with what one might call *a way of viewing things in tune with the dominant theory*, including sometimes even a way of viewing the world and a way of life. Accordingly, such a framework constitutes a social bond between its devotees: it binds them together, very much as a church does, or a political or artistic creed, or an ideology.

This is a further explanation of the asserted incommensurability: it is understandable that *two ways of life and two ways of looking at the world* are incommensurable. Yet I want to stress that *two theories* which try to solve the same family of problems, including their offspring (their problem children), need *not* be incommensurable, and that in science, as opposed to religion, it is the *theories* that are paramount. I do not wish to deny that there is such a thing as a 'scientific approach', or a scientific 'way of life': that is, the way of life of those men devoted to science. On the contrary, I assert that the scientific way of life involves a burning interest in objective scientific theories — in the theories in themselves, and in the problem of their truth, or their nearness to truth. And this interest is a *critical* interest, an *argumenta-*

tive interest. Thus it does not, like some other creeds, produce anything like the described 'incommensurability'.

It seems to me that many counterexamples exist to the theory of the history of science that I have just discussed. There are, first, counterexamples that show that the existence of a 'framework', and of work going on within it, does not characterize science. Philosophy during the scholastic period, astrology, and theology, are such counterexamples. Secondly, there are counterexamples that show that there may be several dominant theories struggling for supremacy in a science, and there may even be fruitful discussions between them. My main counterexample under this heading is the theory of the constitution of matter, in which atomism and continuity theories were, fruitfully, at war from the Pythagoreans and Parmenides, Democritus and Plato, to Heisenberg and Schrödinger. I do not think that this war can be described as falling into the prehistory of science, or into the history of prescience. Another counterexample of this second kind is constituted by the theories of heat. Even after Black we have fluidum theories [16] of heat warring with kinetic and phenomenological theories; and the clash between Ernst Mach and Max Planck [17] was neither characteristic of a crisis nor did it occur within one framework, nor, indeed, could it be described as prescientific. Another example is the clash between Cantor and his critics (especially Kronecker) which was later continued in the form of exchanges between Russell and Poincaré, Hilbert and Brouwer. By 1925 there were at least three sharply opposed frameworks involved, divided by chasms far too wide for bridging. But the discussions continued, and they slowly changed their character. By now not only have fruitful discussions occurred but so many syntheses that the animadversions of the past are almost forgotten. Thirdly, there are counterexamples that show that fruitful rational discussions may continue between devotees of a newly established dominant theory and unconvinced sceptics. Such is Galileo's *Two Principal Systems*; such are some of Einstein's 'popular' writings, or the important criticism of Einstein's principle of covariance voiced by E. Kretschmann (1917), or the criticism of Einstein's General Theory recently voiced by Dicke; and such are Einstein's famous discussions with Bohr. It would be quite incorrect to say that the latter were not fruitful, for not only did Bohr claim that they much improved his understanding of quantum mechanics, but they led to the famous paper of Einstein, Podolsky, and

Rosen which has produced a whole literature of considerable signifi-cance, and may yet lead to more:[18] no paper which is discussed by recognized experts for thirty-five years can be denied its scientific status and significance, but this paper was, surely, criticizing (from the outside) the whole framework which had been established by the revolution of 1925—26. Opposition to this framework — the Copen-hagen framework — is continued by a minority to which for example de Broglie, Bohm, Landé, and Vigier belong — apart from those names mentioned in the preceding footnote.[19]

Thus discussions may go on all the time; and although there are always attempts to transform the society of scientists into a closed society, these attempts have not succeeded. In my opinion they would be fatal for science.

The proponents of the view of the myth of the framework distinguish sharply between rational periods of science conducted within a frame-work (which can be described as periods of closed or authoritarian science) and periods of crisis and revolution, which can be described as the almost irrational leap (comparable to a religious conversion) from one framework to another.

No doubt there are such irrational leaps, such conversions, as described. No doubt there are even scientists who just follow the lead of others, or give way to social pressure, and accept a new theory as a new faith because the experts, the authorities, have accepted it. I admit, regretfully, that there are fashions in science, and that there is also social pressure.

I even admit that the day may come when the social community of scientists will consist mainly or exclusively of scientists who uncritically accept a ruling dogma. They will normally be swayed by fashions; they will accept a theory because it is the latest cry, and because they fear to be regarded as laggards.

I assert, however, that this will be the end of science as we know it — the end of the tradition created by Thales and Anaximander and redis-covered by Galileo. As long as science is the search for truth it will be the rational, critical discussion between competing theories, and the rational critical discussion of the revolutionary theory. This discussion decides whether or not the new theory is to be regarded as better than the old theory: that is, whether or not it is to be regarded as a step towards the truth.

XII

Almost forty years ago I stressed that even observations, and reports of observations, are under the sway of theories or, if you like, under the sway of a framework. Indeed, there is no such thing as an uninterpreted observation, an observation which is not theory-impregnated. In fact, our very eyes and ears are the result of evolutionary adaptations — that is, of the method of trial and error corresponding to the method of conjectures and refutations. Both methods are adjustments to environmental regularities. A simple example will show that ordinary visual experiences have a pre-Parmenidian absolute sense of up and down built into them — a sense which is no doubt genetically based. The example is this. A square standing on one of its sides looks to all of us a different figure from a square standing on one of its corners. There is a real *Gestalt* switch in moving from one figure to the other.

But I assert that the fact that observations are theory-impregnated does not lead to incommensurability between either observations or theories. For the old observations can be consciously reinterpreted: we can learn that the two squares are different positions of the same square. This is made even easier just because of the genetically based interpretations: no doubt we understand each other so well partly because we share so many physiological mechanisms which are built into our genetic system.

Yet I assert that it is possible for us to transcend even our genetically based physiology. This we do by the critical method. We can understand even a bit of the language of the bees. Admittedly, this understanding is conjectural and rudimentary. But almost all understanding is conjectural, and the deciphering of a new language is always rudimentary to start with.

It is the method of science, the method of critical discussion, which makes it possible for us to transcend not only our culturally acquired but even our inborn frameworks. This method has made us transcend not only our senses but also our partly innate tendency to regard the world as a universe of identifiable things and their properties. Ever since Heraclitus there have been revolutionaries who have told us that the world consists of processes, and that things are things only in appearance: in reality they are processes. This shows how critical thought can challenge and transcend a framework even if it is rooted not only in our conventional language but in our genetics — in what may be called human nature itself. Yet even this revolution does not produce a theory incommensurable with its predecessor: the very task of the revolution was to explain the old category of thing-hood by a theory of greater depth.

XIII

I may perhaps also mention that there is a very special form of the myth of the framework which is particularly widespread. It is the view that, before discussion, we should agree on our vocabulary — perhaps by 'defining our terms'.

I have criticized this view on various occasions and I do not have space to do so again.[20] I only wish to make clear that there are the strongest possible reasons against this view; all definitions, so-called 'operational definitions' included, can only shift the problem of the meaning of the term in question to the defining terms; thus the demand for definitions leads to an infinite regress unless we admit so-called 'primitive' terms, that is, *undefined* terms. But these are as a rule no less problematic than most of the defined terms.

XIV

In the last section of this paper I will briefly discuss the myth of the framework from a logical point of view: I will attempt something like a logical diagnosis of the malaise.[21]

The myth of the framework is clearly the same as the doctrine that one cannot rationally discuss anything that is *fundamental*; or that a rational discussion of *principles* is impossible.

This doctrine is, logically, an outcome of the mistaken view that all rational discussion must start from some *principles* or, as they are often called, *axioms*, which in their turn must be accepted dogmatically if we wish to avoid an infinite regress — a regress due to the alleged fact that when rationally discussing the validity of our principles or axioms we must again appeal to principles or axioms.

Usually those who have seen this situation either insist dogmatically upon the truth of a framework of principles or axioms, or they become relativists: they say that there are different frameworks and that there is no rational discussion possible between them, and thus no rational choice.

But all this is mistaken; for behind it there is the tacit assumption that a rational discussion must have the character of a justification, or of a proof or a demonstration, or of a logical derivation from admitted premises. But the kind of discussion which is going on in the natural sciences might have taught our philosophers that there is also another kind of rational discussion: a critical discussion which does not seek to prove or to justify or to establish a theory, least of all by deriving it from some higher premises, but which tries to test the theory under discussion by finding out whether its *logical consequences* are all acceptable, or whether it has, perhaps, some undesirable consequences.

We thus can logically distinguish between *a mistaken method of criticizing* and *a correct method of criticizing*. The *mistaken method* starts from the question: how can we establish or justify our thesis or our theory? It thereby leads either to dogmatism; or to an infinite regress; or to the relativistic doctrine of rationally incommensurable frameworks. By contrast, the *correct method* of critical discussion starts from the question: what are the *consequences* of our thesis or our theory? Are they all acceptable to us?

Thus it consists in comparing the consequences of different theories (or, if you like, of different frameworks) and tries to find out which of the competing theories or frameworks has consequences that seem preferable to us. It is thus conscious of the fallibility of all our methods, and it tries to replace all our theories by better ones. This is, admittedly, a difficult task, but by no means an impossible one.

To sum up. Frameworks, like languages, may be barriers; but a foreign framework, just like a foreign language, is no absolute barrier. And just as breaking through a language barrier is difficult but very much worth our while, and likely to repay our efforts not only by widening our intellectual horizon but also by offering us much enjoyment, so it is with breaking through the barrier of a framework. A breakthrough of this kind is a discovery for us, and it may be one for science.

Penn, Buckinghamshire   KARL POPPER
England
October, 1972

NOTES

* Based on a paper which I first prepared in 1965. I am indebted to Arne Petersen and Jeremy Shearmur for various suggestions and corrections. The motto is from Plato's *Crito*, 49D.

From: *The Abdication of Philosophy: Philosophy and the Public Good*. Essays in Honor of Paul Arthur Schilpp. Edited by Eugene Freeman. La Salle, Illinois: Open Court, 1976, pp. 23—48.

[1] Herodotus, III, 38. I refer to this passage in n. 3 to Chap. 5 of my *Open Society and Its Enemies*. London: George Routledge & Sons, 1945; Princeton: Princeton University Press, 5th rev. ed., 1966. Vol. I.

[2] The distinction between nature and convention is discussed in my *Open Society*, Vol. I, Chap. 5, where I refer to Pindar, Herodotus, Protagoras, Antiphon, Archelaus, and especially to Plato's *Laws* (cp. nn. 3, 7, 10, 11, and 28 to Chap. 5 and text). Although I mention (p. 60) the significance of 'the realization that taboos are different in various tribes', and although I (just) mention Xenophanes (n. 7) and his profession as a 'wandering bard' (n. 9 to Chap. 10), I did not then fully realize the part played by culture clash in the evolution of critical thought, as witnessed by the contribution made by Xenophanes, Heraclitus, and Parmenides (see esp. n. 11 to the *Open Society*, Chap. 5) to the problem of nature or reality or truth versus convention or opinion. See also my *Conjectures and Refutations: The Growth of Scientific Knowledge*. New York: Basic Books, 1963; London: Routledge & Kegan Paul, 4th rev. ed., 1972. Passim.

[3] Cp. my *Conjectures and Refutations*, 4th rev. ed., pp. 152 f. The first two lines of my text are fragment B 16 and the next four fragment B 15. The remaining three fragments are B 18, 35, and 34 (according to Diels-Kranz, *Fragmente der Vorsokratiker*. 5th ed.) The translations are mine. Note, in the last quoted two lines, the contrast between the one final truth and the many guesses, or opinions, or conjectures.

[4] Parmenides used Xenophanes' terminology; see *Conjectures and Refutations*. 4th rev. ed., e.g., pp. 11, 17, 145, 400, 410. See also my *Open Society*. Vol. I. n. 56, section (8). to Chap. 10. p. 312.

[5] See Parmenides' remark (in fragment B 6) on the muddled horde of erring mortals, always in two minds about things, in contrast with the one 'well rounded truth'. Cp. *Conjectures and Refutations*, pp. 11, 164 f.

[6] *Theogony*, 720—25.

[7] *Iliad*, VIII, 13—16; cp. *Aeneid* VI, 577.

[8] See my *Conjectures and Refutations*, 4th rev. ed., pp. 126 ff. 138 f., 150 f., 413.

[9] *Theogony*, 720—25.

[10] The discovery is, it appears, due to Parmenides; see fragments B 14—15:

> Bright'ning the night she glides round the earth with a light that is borrowed;
> Always she wistfully looks round for the rays of the sun.

[11] See Alfred Tarski, *Logic, Semantics, Metamathematics*, trans. by J. H. Woodger. New York: Oxford University Press, 1956. I have expounded it in various places; see, for example, my *Conjectures and Refutations*, pp. 223—25.

[12] See Benjamin Lee Whorf, *Language, Thought, and Reality*, ed. by John B. Carroll. Cambridge, Mass.: MIT Press, 1956.

[13] See W. V. Quine. *Word and Object*. Cambridge, Mass.: MIT Press, 1960; and *Ontological Relativity and Other Essays*. New York: Columbia University Press. 1969.

[13a] Cp. p. 232 of T. S. Kuhn, "Reflections on my Critics", in *Criticism and the Growth of Knowledge*, ed. by Imre Lakatos and Alan Musgrave. London: Cambridge University Press, 1970, pp. 231—78.

[14] When writing this section, I had originally Thomas Kuhn in mind, and his book *The Structure of Scientific Revolutions*. Chicago: Chicago University Press, 1962, 1970. (See also my contribution, "Normal Science and its Dangers", to *Criticism and the Growth of Knowledge*, ed. by Imre Lakatos and Alan Musgrave. London: Cambridge University Press, 1970, pp. 51—58.) However, as Kuhn points out, this interpretation was based on a misunderstanding of his views (see his "Reflections on my Critics", in *Criticism and the Growth of Knowledge*, pp. 231—78; and his "Postscript 1969" to the 2nd ed. of *The Structure of Scientific Revolutions*), and I am very ready to accept his correction. Nevertheless, I regard the view here discussed as influential.

[15] For a Criticism of Karl Mannheim, see Chaps. 23 and 24 of my *Open Society*. Vol. II.

[16] Few people seem to realize that by his equation $E = mc^2$, Einstein resurrected the fluidum theory of heat (caloric) for which the question whether heat has any weight was regarded as crucial. According to Einstein's theory, heat *has* weight — only it weights very little.

[17] Cp. the discussion between Planck and Mach, especially Planck's paper "Zur Machschen Theorie der physikalischen Erkenntnis", *Physikalische Zeitschrift* **11** (1910): 1186—90.

[18] See, for example, J. S. Bell, "On the Einstein Podolsky Rosen Paradox", *Physics* **1** (1964): 195—200; J. S. Bell, "On the Problem of Hidden Variables in Quantum Mechanics", *Reviews of Modern Physics* **38** (1966), 447—52; John F. Clauser, Michael A. Horne, Abner Shimony, and Richard A. Holt, "Proposed Experiment to Test Local Hidden Variable Theories", *Physical Review Letters*, October 13, 1969. An extension or strengthening of the EPR paradox described in my *Logic of Scientific Discovery*. New York: Basic Books, 1959, 1972, pp. 446—48, seems to me to involve a decisive refutation of the Copenhagen interpretation since the two simultaneous measurements together would allow simultaneous 'reductions' of the two wave packets which cannot be carried out within the theory. See also the recent paper by James Park and Henry Margenau, 'Simultaneous Measurability in Quantum Theory', *International Journal of Theoretical Physics* **1** (1968): 211—83.

[19] See my paper "Quantum Mechanics Without 'The Observer'", in *Studies in the Foundations, Methodology and Philosophy of Science*, Vol. 2: *Theory and Reality*, ed. by Mario Bunge. New York: Springer-Verlag, 1967.

[20] See my *Open Society*, Vol. II, Chap. 11, Sec. II; or my paper "Quantum Mechanics without 'The Observer'", esp. pp. 11—15; or my *Conjectures and Refutations*, pp. 19, 28, section (9), and pp. 279, 402.

[21] I am greatly indebted to my friend Alan Musgrave for reminding me to include in this paper the logical diagnosis contained in the present section.

JOHN WATKINS

A NEW VIEW OF SCIENTIFIC RATIONALITY

No one can doubt the extraordinary technological power that modern science has given to the western world. As Bertrand Russell put it:

> Science, as a dominant factor in determining the beliefs of educated men, has existed for about 300 years; as a source of economic technique, for about 150 years. In this brief period it has proved itself an incredibly powerful revolutionary force.[1]

And even if we set aside its industrial, medical and military applications, and consider science as a purely theoretical system, a systematic attempt to describe and explain the workings of the world we live in, it is still extraordinarily impressive. It is as if, while the rest of mankind had built only mud-huts, a comparatively small number of people in a comparatively short time had built a soaring cathedral.

But when we raise philosophical questions about the cognitive status of the theories of modern science, we get no clear, cogent, positive, and generally agreed answers. Can these theories be known to be true? There was a time when it was widely believed that they could be. If a theory, such as Newton's, yielded a great variety of striking predictions, all of which were subsequently verified, then surely it *must* be true? But this confident appraisal suffered two hammer blows, as we shall see. Then can a good scientific theory, though not certainly true, at least be known to have a high probability of being true? Or if it cannot attain a high probability, can the mass of evidence that bears it out at least *raise* its probability? Although there are philosophers who answer this last question affirmatively, the case for a negative answer is, I believe, overwhelming. Then can we at least know of a successful theory, that has superseded a less successful one, that although the former's probability has not been raised and is perhaps stuck at zero, it is closer to the truth that the latter? Again the answer has to be no. There is at the present time a vociferous group who hate and despise modern science; and to them these negative results are most welcome; for they seem to imply that science is just another irrational ideology, on a par with, say, Zande magic. But to those of us who admire science as one of the outstanding achievements of western civilization, these results constitute an urgent challenge.

63

J. C. Pitt and M. Pera (eds.), Rational Changes in Science, 63–82.
© *1987 by D. Reidel Publishing Company.*

It is ironic that in the 'glad confident morning' of modern science, in the early seventeenth century, when it had made only rather small beginnings, some philosophers were filled with a shining optimism as to the certainty and ultimacy of the theoretical understanding of the world that science could and would achieve in the not too distant future; whereas today there is widespread philosophical disillusion and scepticism concerning the cognitive status of science, despite the enormous strides it has taken since those early days. In Section I we will look into the historical reasons for this curiously inverse correlation. In Section II, a way of restoring philosophical confidence in science will be indicated.

I

In the early seventeenth century there were two great visionary spokesmen for the new science: Francis Bacon and René Descartes. There were large differences between their views of the way in which science should be advanced. But what is striking is the similarities between their visions of what science would achieve. Both men believed that the human understanding, properly regulated, can get to the very bottom of things, unlock Nature's deepest secrets, grasp her ultimate essences. And they both believed that the knowledge to be acquired at this ultimate level could be certain or infallible. We could summarise this 'Bacon-Descartes ideal', as I call it, as the claim that science can arrive at *ultimate explanations* that are *certainly true*.

There was a period when it seemed that something approaching the Bacon-Descartes ideal had actually been achieved. When Isaac Newton published his *Principia* in 1687 it did not immediately win general acceptance. There was resistance on the Continent from followers of Descartes who objected to one of its central ideas, and one which Newton himself found perplexing, namely that two bodies, however widely separated by empty space, act instantaneously upon each other in accordance with the law of universal gravitational attraction; again, Newton's ideas of absolute space and time ran into sharp opposition from Leibniz, Berkeley and others; and at the empirical level the erratic ways of the moon gave Newton endless trouble. But as the years passed, empirical difficulties of this latter kind were overcome, one after another; the theory enjoyed a succession of brilliant predictive successes (concerning, for instance, the perturbations of Saturn and

Jupiter when in conjuction, the return of Halley's comet in 1759, which Halley had predicted back in 1682, and the discovery of Neptune, from calculations based on the "misbehaviour" of Uranus, in 1846). As to the theoretical difficulties: the idea of gravitational attraction between bodies at a distance lost its strangeness and began to seem plausible; indeed, that the attraction must vary inversely with the square of the distance between the bodies now seemed to be dictated by geometrical considerations. Surely this astonishingly successful theory of heavenly and terrestrial mechanics was *true*. And did it not reveal, if not God's entire plan for His creation (for it did not cover electrical, chemical, and biological phenomena), a major part of it?

And then, in 1739, David Hume published Book I of his *Treatise of Human Nature*. Little notice was taken of it at first; it lay around for a quarter of a century, like a time-bomb quietly ticking away, before attention was first drawn to the danger, by Thomas Reid in his *Inquiry into the Human Mind*, 1764. Tucked away in Hume's sprawling work are two propositions which, between them, seem to destroy the possibility of genuine scientific *knowledge*. One is that no matter of fact can be settled by reason unaided by experience: anything that we can conceive at all we can conceive as existing, or as not existing; whether it does in fact exist can be decided only by experience. The other is that neither reason nor experience can, in Hume's words, prove *that those instances, of which we have had no experience, resemble those, of which we have had experience*. If reason and experience together cannot even show that the sun will rise tomorrow, or that the next glass of water a thirsty man drinks will quench his thirst, then they obviously cannot show that a grand scientific theory such as Newton's, with its endless array of predictive implications for the future, is true.

'Answering Hume' has developed into a philosophical industry. After Reid there came Oswald, Beattie, and Priestley, and then the greatest Hume-answerer of all, Immanuel Kant. Kant said of his predecessors, "One cannot observe without feeling a certain pain, how his [Hume's] opponents Reid, Oswald, Beattie and finally Priestley, so entirely missed the point of his problem".[2] Kant was entirely persuaded of the truth of Newtonian mechanics. He was also persuaded by Hume that such a soaring theoretical edifice could not be supported merely by a combination of experience and logic. Something more was needed, and Kant believed that he could supply this extra something. If we are to understand his answer to Hume we need to bear in mind two important differences between mathematics in his day and in ours. First, non-Euclidean geometries had not yet been developed: Euclidean geometry reigned supreme and was accepted as *the* theory of space, necessarily true and knowable *a priori*. Second, Kant saw pure mathematics as an autonomous system: he would have regarded our twentieth century attempts to reduce mathematics to logic as fundamentally mistaken; for logic is analytic whereas mathematics, Kant insisted, is synthetic. An analytic truth is empty, has no factual content. But mathematical truths, Kant insisted, are not empty. They have content. However, their content is not empirical; no experience could conflict with them, nor does their content derive from experience. They have a *pure* content derived not from sensory intuitions (say of coloured surfaces) but from our pure intuitions of space and of time. (Pure intuitions of space supply geometry with its content, while pure intuitions of time provide an idea that is indispensable for arithmetic and the construction of numbers, namely the idea of *succession*.)

Another vital distinction drawn by Kant was of course that between truths that are *a priori* and truths that are *a posteriori*. This distinction concerns the way in which a truth can be known. An *a priori* proposition can be known to be true by reason independently of experience: it is necessarily true. An *a posteriori* truth can be known only from experience, and is only contingently true. Now mathematical truths are not contingent truths known from experience: their truth is necessary and can be known *a priori*. In short, mathematics consists of synthetic *a priori* truths.

But mathematics, though *a* stronghold of synthetic *a priori* truths, was not, for Kant, their only domain: there is such a thing as Pure Natural Science which likewise consists of synthetic *a priori* truths, of which Kant claimed to have provided a complete inventory and infallible proofs. And armed with these proofs Kant could answer Hume: if, as Hume supposed, our scientific knowledge was constructed only from experience with the help of mere logic, it would collapse; but it does not collapse because its construction involves, in addition, a steely and immutable framework of synthetic *a priori* principles.

Among the principles which he included in this category were: the permanence of substance, the law of causality, and the principle that all substances coexisting in space are in thoroughgoing reciprocity. These bear an obvious resemblance to certain key principles within classical Newtonian mechanics, namely the conservation of matter, physical determinism, and universal gravitational attraction. Unfortunately for Kant, the rapid and explosive progress of physics since his time has burst asunder that steely framework: the creation of matter and physical indeterminacies are admitted by modern physics, and action-at-a-distance is disallowed. Kant's system is a magnificent ruin and the apriorist answer to Hume is defunct.

When it came to be recognized that science cannot achieve certainty, it was widely accepted that it can, however, achieve something approximating certainty, namely high probability. Moritz Schlick was one of many philosophers who took this view: the truth of scientific hypotheses is not absolutely guaranteed, he said; we must be content if their probability is extremely high.[3] *Probabilism*, as we may call this approach to Hume's problem, is at the present time the biggest and busiest branch of the Hume-answering industry. One of its inaugurators was Bolzano, whose *Wissenschaftslehre* was published in 1837. It has been carried on by Jevons, Keynes, Jeffreys, Carnap, and many more recent thinkers.

Bolzano's idea was as follows. If evidence e entails a hypothesis h then h holds in every possible world in which e holds and the probability of h given e is one, or $p(h, e) = 1$. Conversely, if e contradicts h then h holds in no possible world in which e holds and $p(h, e) = 0$. If h holds in *nearly all* the possible worlds in which e holds, then the probability of h given e is close to one. Again, if h holds in *half* the possible worlds in which e holds, then $p(h, e) = 1/2$.[4]

This idea obviously created the need to find a precise representation of the vague idea of a 'possible world'. Carnap met this need with his concept of a *state-description*.[5] Given a suitable and well-specified language, a state-description characterizes every individual in its domain as completely as the language allows. Actually, Carnap chose as the basic units for his probability logic, not state-descriptions but structure-descriptions. Two state-descriptions are isomorphic, or have the same structure, if one can be got from the other merely by permuting two or more of the individual constants; and a structure-description creams off the common content of all those state-descriptions that are isomorphic with one another.[6] Hintikka chose as the basic units for his system not structure-descriptions but constituents.[7] A structure-description does not say which individuals instantiate a certain predicate but it does say how many anonymous individuals instantiate it. A constituent (more precisely, a constituent of depth-1) does not say how many individuals instantiate it but only that this predicate is instantiated by at least one anonymous individual. It creams off all the qualitative content of a structure-description. An important advantage, from a probabilist point of view, of a constituent over a structure-description is that the initial probability of the former, being independent of the number of individuals in the domain, may remain positive even if the domain is infinite, whereas the initial probability of the latter tends to zero as the number of individuals tends to infinity.

The most influential version of probabilism at the present time is Bayesianism. Let us write $p(h, e)$ to denote the posterior probability of a hypothesis h on evidence e and $p(h)$ to denote the probability of h prior to the arrival of evidence. The distinctive feature of Bayesianism is, of course, that it is primarily interested, not in the absolute level of $p(h, e)$, but in the *ratio* of the value of $p(h, e)$ to that of $p(h)$. Suppose that $p(h, e) = 0.01$. Then if $p(h) = 0.0001$ the evidence e will have raised the probability of h a hundredfold and will have strongly confirmed it, even though the absolute level of $p(h, e)$ is low. Conversely, if $p(h) = 0.1$ then e will have disconfirmed h. If $p(h, e) = p(h)$ then e is neutral to h.

However Bayesianism does not avoid some severe difficulties which confront any version of probabilism. One of these is connected with the fact that if the prior probability of h is zero, then no evidence, however seemingly favorable to h, can raise its posterior probability above zero. Now the laws of physics are precise and universal: they apply to all space-time regions. And in nearly all systems of probability logic (Hintikka's is an exception) the probability of such a statement is always zero. Various attempts have been made to get around this major difficulty. Jeffreys tried to overcome it with the help of his simplicity-postulate.[8] After providing an admirably clearcut definition of simplicity, he postulated that the simplest of a set of alternative hypotheses all compatible with the given evidence is, other things being equal, the one that is the most probable. But it turned out that his simplicity-postulate leads to inconsistencies, as Popper pointed out.[9] The underlying reason for this is that the simpler, in Jeffreys's sense, a hypothesis is, the stronger it is or the more it asserts; but it is a paramount principle of probability logic that if one hypothesis says *more* than another then, other things being equal, it has a *lower* probability than the other (or at any rate not a higher probability; the probability of both hypotheses may be zero).

Hintikka developed a system in which universal laws can have a positive probability; for the probability of a law is the sum of the probabilities of each of the constituents that entail it; and these constituents can have positive probabilities, for the reason indicated above. And if a sufficient amount of favorable evidence comes in, the probability of a law can rise to a high level. But his system involves something analogous to Jeffreys's simplicity-postulate: a time comes when it is the *strongest* of all the unrefuted constituents that is treated as the most probable, contrary to the paramount principle mentioned above. It should be the weakest unrefuted constituent that has the highest probability. Now this constituent is the one that asserts *every existential statement* expressible in the language, and hence *denies* every law-statement. And as evidence comes in, a higher proportion of the probabilities of the refuted constituents should accrue to this weakest constituent than to unrefuted constituents that are stronger than it. Hence the impact of increasing evidence on the law, no matter how favourable it may seem, should be to *lower* (or at any rate, not to raise) its probability.

Another difficulty for probabilism, which I consider insuperable, is the following. Let E be some considerable body of evidence and suppose that $p(h, e)$ is much higher than $p(h)$. But before we can conclude that h is very well confirmed we need some assurances concerning E. One is that E does not contain misinformation: all the evidence-statements comprised in it should be verified. And many philosopers (myself included) hold that that is not possible. However that is not the difficulty to which I now wish to draw attention. Suppose for the sake of the argument that E *is* known to be true. We still need a further assurance. For suppose that there were a bit of evidence in our possession that is not included in E and which actually refutes h. Then the fact that $p(h, E)$ is high (or higher than $p(h)$) is unimportant: the important fact is that the probability of h on *all* the evidence in our possession is zero. In other words, we need to know that E contains (i) *only* evidence-statements that are known by us at the present time to be true and (ii) *all* such statements. This is the famous Requirement of Total Evidence. Unless we can be assured that E satisfies this requirement we are not entitled to base an appraisal of h on the mere fact that $p(h, E)$ is high.

But now consider whether a person X could ever know, of a given E, that it satisfies this requirement. To make X's task easier, assume (counter-factually) that, confronted by a particular evidence-statement,

X can tell unhesitatingly whether or not he knows it to be true, and that he checks through every evidence-statement contained in E and finds in each case that he does indeed know it to be true. That establishes (i) that E contains *only*, but not (ii) that it contains *all*, evidence-statements known to him. It is as if, to vary Descartes's simile, he needed to establish that a certain basket of apples contains (i) none that are rotten and (ii) all that are not rotten. To verify that E satisfies the total evidence requirement X would need to check that everything outside E is *not* evidence known to him. And the situation becomes even worse if the requirement is modified to one of total *relevant* evidence. To verify the claim that E satisfies this latter requirement, he would need to check that everything inside E is both known to him and relevant to the hypothesis in question, and *also* that everything outside E is not known to him or, if known, is not relevant to the hypothesis. And it is obvious that even if the first part of this task could be completed, the second part could not.

Hume's problem could be restated as follows: inductive inferences (that is, inferences from observed to unobserved instances) are (1) *indispensable* (both in everyday life and in science) and (2) *unjustifiable* (neither reason nor experience can provide any justification for them). Hume himself recommended 'carelessness and inattention' as the only remedy for this painful dilemma.[10] We might say that on this view (a softer version of which has been taken up by Strawson and Ayer in more recent times) the "rationality" both of common sense and of science lies in the fact that the human mind operates in a robustly non-logical way.

So Hume retained both (1) and (2), despite their mutual antagonism. Both Kant and the probabilists retained (1) and rejected (2). Kant rejected (2) on the ground that logic and experience are reinforced by a framework of synthetic *a priori* principles; probabilists rejected (2) on the ground that classical deductive logic can be generalized and extended to probability logic within which inductive inferences can be accommodated. As we have seen, these two answers fail; and Hume's 'answer' is really no answer. A remaining possibility is to retain (2) and reject (1). This is what Karl Popper did.[11] According to him there are, in science, no inferences from evidence to laws and theories. The latter remain forever conjectural hypotheses, and all scientific inferences are deductive. From these conjectural theories are deductively derived various predictive consequences; and the latter, especially the more novel and unexpected ones among them, are put to the test. If a test-result is negative there is a deductive inference from this to the falsity of the theory under test. If a theory has stood up to severe testing it is said to be highly *corroborated*; but this does not mean that it is very probably true or anything of that sort; it means only that the theory has performed very well so far. The chief rule of Popper's methodology is that we should provisionally adopt, as the best theory in its field, the one that is best corroborated.

An important advantage of this rule, compared with one based on probabilities, is that it is not prejudiced against powerful scientific theories. On the contrary, the more powerful a theory is the more corroborable it is; and the most corroborable of a set of competing theories may go on to become the best corroborated of them, if it survives severe testing. But the question remains as to why, from Popper's anti-inductivist point of view, which prohibits inferences from evidence to the probable truth of a theory, we should regard the best *corroborated* of a set of competing theories as being, at present, the *best* theory in the set. Were it the sole unfalsified survivor in the set, we would have a good reason to prefer it. But there are bound to be other unrefuted hypotheses: for instance, weaker ones that are strictly entailed by the best corroborated theory. These will be, typically, less corroborable and less well corroborated than it. They will also be 'safer' in the sense that they are less likely to be overthrown by future tests. Why should we prefer it to them?

Popper's main answer has been that if one theory is better corrobo-rated than another, then "we will, in general, have *reason to believe* that the first is a better approximation to the truth than the second".[12] In other words, we have reason to believe that the first has the greater verisimilitude. But this answer involves a lurch into inductivism. Consider two hypotheses, A and B, which diverge but which have equal amounts of content, in that the consequences of A are in one: one correspondence with those of B. (I call such hypotheses 'incongruent counterparts'.) Then the verisimilitude-appraisal that A is closer to the truth than B carries the implication that, if a crucial experiment between them, of a hitherto untried kind, is carried out in the future, then A is more likely to pass this test than is B. For we can think of the predictive implications of A and B on the analogy of two urns containing equal numbers of balls, some white (= true), the others black (= false), and with urn A containing a higher proportion of white balls; and a crucial experiment, that is, an experiment directed at certain divergent consequences of A and B, is analogous to selecting equal numbers of balls from the two urns. It is more probable that those selected from urn A will be white than that those selected from B will be. Now a corroboration-appraisal, Popper has insisted, is "an evaluating *report of past performance* ... [and] *says nothing whatever about future performance*".[13] But verisimilitude-appraisals, as we have just seen, *do* carry predictive implications as to future performance. Hence if corroboration-appraisals are taken to provide some sort of justification for verisimilitude-appraisals they *will* be saying something indirectly, or at one remove, about future performance.

II

Our conclusions so far have all been negative. Philosophical confidence in science will hardly be restored if it essentially involves invalid inferences *a la* Hume, Ayer[14] and Strawson.[15] The Kantian idea of stiffening science with a framework of synthetic *a priori* principles has collapsed. The probabilist idea that it is rational for a person X to prefer, given a set of competing hypotheses, the one that is best confirmed, in some probabilist sense, by the total evidence turns out to be unworkable. And the justification for Popper's idea that it is rational to prefer the one that is best corroborated, in his non-probabilist and non-inductivist sense, turns out to have a tacitly inductivist character: there is a slide from, in Humean language, instances of which we have had experience (past performance) to expectations about instances of which we have had no experience (future performance).

    Then has the idea of scientific rationality broken down? Are there never any good cognitive reasons for accepting a scientific theory as the best one in its field at the present time? It is to this problem that my *Science and Scepticism*[16] is addressed. This book is in the Popperian tradition, but it departs from his views in various ways. The idea of one theory being closer to the truth, or having more verisimilitude, than another plays no role in it. A new theory of the rational, or at least quasi-rational, acceptance of statements into the empirical basis is provided. (Briefly, the idea is that a perceptual experience is treated, not as a premise from which a basic statement about the external world is inferred, but as an explanandum for which a conjectural explanans has been found in which the basic statement figures essentially.) Another difference concerns the comparative measure of the testable content of two theories. It turned out that none of Popper's measures could handle the case where theory *A* goes beyond *and revises* theory *B*, a case which we very much need to be able to handle, since theories that satisfy Bohr's Principle of Correspondence are of that kind. I dealt with this difficulty by generalizing Popper's original measure, which relied on the subclass relation between the theories' classes of potential falsifiers, with the help of the idea, touched on above, of *incongruent counterparts*. Since their consequences are in one: one correspondence, incongruent counterparts have equal amounts of testable content; and I say that *A* has more testable content than *B* if there exists a counterpart *B'* of *B* such that every potential falsifier of *B'* is a potential falsifier of *A* but not vice versa. (This allows that *B'* may be a *congruent* counterpart of *B*, or logically equivalent to *B*.)

Concerning the central problem of the rational acceptance of scientific hypotheses, my idea was this. Suppose that we could establish that a certain aim is *the optimum aim* for science; and suppose, further, that we could establish that the best corroborated theory in its field, if there is just one, is the one, at the present time, that best fulfills this aim. Then we would have the best possible reason to accept this theory.

But how can we possibly arrive at the optimum aim for science? I began by laying down certain obvious adequacy requirements for any aim for science. The two most important ones are that the aim must be *coherent* (it must not contain diverse components that pull in opposite directions) and *feasible* (it must be capable of being fulfilled). The question then becomes: is there a maximumly ambitious aim that conforms with these adequacy-requirements and which would no longer do so if it were further strengthened? If there is, then this aim would *dominate* any other aim that conforms with them, since it would contain an aspiration not contained in the other one but not vice versa.

My starting point in the search for such a dominating aim was what I call the Bacon-Descartes ideal, which I interpret as saying that it is the goal of science to render all phenomena explainable and predictable by rigorous deduction from (appropriate initial conditions and) universal principles that are certainly true, ultimate, unified, and exact. It seems obvious that, as a first step in attempting to turn this into a feasible ideal, it should be recast as the aim of *progressing towards*, rather than actually attaining, these far-off and perhaps unattainable goals. Thus "progressivised", the aim of science becomes to advance with explanatory theories that are ever more:

(A)   probable;
(B1)   deep;
(B2)   unified;
(B3)   predictively powerful;
(B4)   exact.

I call the (A)-component of this version of the Bacon-Descartes ideal its security-pole and the (B)-components its depth-pole. As we shall see, it is not, as it stands, a coherent aim of science because its two poles pull in opposite directions. Let us start by considering its depth-pole.

I will not reproduce here the rather technical elucidations given in my book of the ideas of one theory being deeper, and more unified than another, though I will say something later about the main idea behind them. And I may mention that they were governed by what I call an *anti-trivialization* principle. This says that if a philosophical account of some aspect of science has the (no doubt unintended) implication that it is trivially easy to create a bit of scientific "progress" (for instance, by plugging into an existing theory a few more theoretical predicates that do no empirical work, or by tacking on to it an idle metaphysical conjunct), then there must be something wrong with that account. Actually, it turns out that the conditions for one theory to be deeper are essentially the same as those for it to be more unified than another. Instead of two distinct components, (B1) and (B2), we really have one component, call it (B1—2). As to (B3), the demand for greater predictive power, it obviously calls for excess testable content. And the same is true of (B4), or the demand for increasing exactitude. Actually, (B3) and (B4) are not really two distinct components, but merge into one component, call it (B3—4), just as (B1) and (B2) merged into (B1—2). For a universal theory can always be restated in such a way that it ascribes *one* predicate (which may be very complex) to *everything*; and we can say that one empirical theory is more exact than another if, when so restated, the predicate it ascribes is more exact than the one the other ascribes to everything. But in that case the former theory will be more testable and have greater predictive power.

Thus all the (B)-components of our progressivised version of the Bacon-Descartes ideal are calling, directly or indirectly, for increasing testable content as science progresses. Now consider component (A). A

fundamental negative result obtained by Popper[17] was that the probability of hypotheses, other things being equal, tends to vary *inversely*
with their testability. (I inserted the words 'tends to' to allow for the
possibility that two hypotheses, one of which is more testable than the
other, both have zero probability.) If increasing probability were our
aim, we should prefer some weak consequence of a highly testable
theory to the theory itself, since there is bound to be a consequence
that is more probable than the theory. Thus the security-pole and the
depth-pole of the Bacon-Descartes ideal pull in opposite directions, one
calling for less, and the other for more, testable content.

If one compares what was happening *in* science from, say, Planck's
discovery of the quantization of energy around 1900 to the discovery
of nuclear fission in the 1930s, with what such philosophers as Mach,
Duhem, Bridgman, and the members of the Vienna Circle were saying
about science, one is struck by a remarkable contrast. It is as if Planck,
Einstein, Rutherford, Bohr and others were inspired by the depth-pole
of our ideal to penetrate to ever deeper layers of physical reality,
whereas those philosophers of science were demanding that scientists
should eschew depth and remain on the surface, at the phenomenal
level. No doubt the various philosophers who waged this 'antidepth
war', as Mario Bunge called it, [18] each had his own set of reasons which
differed somewhat from one to another. But my own diagnosis is that
they intuitively perceived the antagonism between depth and security,
and renounced the former in the hope of attaining the latter. This is
particularly clear in the case of Moritz Schlick. In his *Allgemeine
Erkenntnislehre* (1918) he had not yet perceived the antagonism
between (A) and (B) and he strongly endorsed them both. As to (A): he
spoke of the 'requirements for rigor and certainty' in science, of the
'goal of absolute certainty and precision'; but he also conceded, as we
saw earlier, that the truth of scientific hypotheses 'is not absolutely
guaranteed', adding that we must be content if their probability
'assumes an extremely high value'. As to (B): this book is characterized
by a strong scientific realism: science penetrates beneath the surface to
underlying realities. It tells us 'about the interior of the sun, about
electrons, about magnetic field strengths'. Indeed, science gets to the
very essences of things: 'Maxwell's equations disclose to us the
"essence" of electricity, Einstein's equations the essence of gravitation'.
But by 1931 Schlick had become acutely aware of the conflict between
the goal of high probability and the goal of explanatory depth; and

he was anxious to solve Hume's problem. He continued to adhere resolutely to (A) and now totally repudiated (B): the 'laws' of science were now understood, not as genuine assertions about the world but as rules that license inferences from singular observation statements to other singular observation statements. (For references see my *Science and Scepticism*.)

Now if both the (A)-component and the (B)-components of the Bacon-Descartes ideal constituted possible aims for science which, though mutually exclusive, are separately feasible, then there would be no such thing as the optimum aim for science. We would face a choice between two aims neither of which dominates the other. But (A) is not a feasible aim for science. I mentioned earlier what I consider to be one insuperable difficulty confronting probabilism: a probabilist, whether or not he is a Bayesian, must judge the posterior probability of a hypothesis in the light of the total evidence; but this is something he can never do. I also mentioned earlier arguments for the thesis that evidence, however favorable, can never *raise* the probability of a law-statement. For if, as in many systems of probability logic, including Carnap's, the initial probability of such a statement is zero, then its posterior probability will also always be zero; and if, as in Hintikka's system, its initial probability is positive, then for the reason indicated above, evidence can only lower its probability. In *Science and Scepticism* I also go into what I call the "next-instance" thesis, according to which the predictive content of a theory should be taken to be, not the totality of its predictive implications, but only what it predicts for the next instance (or the next application, or the next year, say). Now if 'next' is here used as in 'The next number in the sequence 1, 2, 4, 8, 16 ... is always twice the previous number', it turns into a variable that extends over all future instances, the theory's predictive content has not been cut back, and the previous sceptical arguments continue to apply. Then suppose that 'next' is used as in 'The next item on the agenda is the treasurer's report', so that it becomes a singular denoting term. Understood in this sense, the "next-instance" thesis would indeed effect a drastic reduction in the predictive content of a theory, making it like a bee that can only sting once. Would this drastic measure at least give the probabilist the desired result that evidence can raise the probability of the little that remains of the theory? Carnap had claimed that instead of taking the roundabout way from evidence via the theory to a prediction yielded by it, we should go straight from the evidence to the prediction.[19] Now a good scientific theory makes *novel* predictions; and there must be a time when the next instance is also the *first* instance of such a prediction. An example used by Hilary Putnam, in criticism of Carnap, was the prediction of the first nuclear explosion.[20] In his reply, Carnap conceded that the totality of the evidence available beforehand would not assign 'a considerable value' to the probability of the prediction.[21] But the question is not whether it would raise it to a considerable value, but whether it would raise it at all. And it is clear that it would not. It would be as if the evidence concerned red robins, black crows, white swans, etc., and the prediction were of a remarkable bird of a kind that hitherto had never been observed.

My conclusion in this part of the book is that those antidepth philosophers of science who attempted to propitiate this almost universally accepted aim, gave an increasingly impoverished account of science in which one valuable feature after another (realism, universality) of science was given up, in the hope of rendering it probabilifiable. *And all these sacrifices were in vain*: the emasculated residue was still incapable of attaining the kind of probability that was sought.

If that is so, the question arises whether there is, within component (A), a feasible core that we can retrieve and combine with the (B)-components, thus creating a stronger aim that would dominate any other aim that is coherent and feasible, and hence would be the optimum aim for science. In the original, absolute version, component (A) called for scientific theories that are *certainly true*. Spelt out more fully, it said that a necessary and sufficient condition for it to be rational for person X to accept a scientific theory at time t is that the theory is certainly true for X at t. In its "progressivised" version, it called for scientific theories that are ever more *probably true*. Now if, as I argue in *Science and Scepticism*, a scientific theory, however much de-ontologized and emasculated by positivist reinterpretation, cannot even have its probability *raised* by seemingly favorable evidence, then the next weakening of component (A) that is needed to render it feasible must presumably be to make it call for scientific theories that are *possibly true*. Spelt out more fully it would now say that a necessary, but not a sufficient, condition for it to be rational for a person X to accept a scientific theory at time t is that the theory is possibly true for X at t. But just what would this mean? We might try saying that it means that X knows that the theory is internally consistent and that there are no conflicts between it and the evidence in his possession at t.

But that would again render it an infeasible demand. It is impossible for
X to be aware of all the logical implications of the theory, because
there are infinitely many of these. And it is impossible for X to say of
any large body of evidence E that E is the *total* evidence in his
possession at t, for reasons given earlier. Thus it will always be possible
that there is a consequence of this theory that does in fact conflict with
the evidence in X's possession, though X has not noticed this. Then
should we say that a theory is possibly true for X at t if X does not
know of any internal inconsistencies in it or of any conflicts between it
and the evidence in his possession at t? But that would render it too
lax. Perhaps X is a lazy fellow who prefers not to search for internal
inconsistencies or adverse evidence. If our modified version of (A) is to
be neither too strong nor too weak, then it should say that it is a
necessary condition for the rational acceptance of a scientific theory by
X at t that X has not succeeded in discovering any internal inconsistencies in it, or evidence adverse to it, despite his best endeavors to
do so.

Combining this modified version of (A) with the previous (B)-
components results in the normative claim that, of all the competing
scientific theories in a certain field that are possibly true in the above
sense, the best is the one that is deepest and most unified, predictively
powerful and exact. The question now is whether, if a certain theory is
the best *corroborated* one in its set, it is thereby the *best* theory in the
above sense.

My account of corroboration is derived from Popper's, though it
differs over some details. In his theory, so-called "background knowledge" plays an essential role, since the measure of the severity of a test
is given by $p(e, hb) - p(e, b)$, where e is a predicted test-result, and b
is background knowledge. Now if $p(e, b)$ is to have a determinate value
we need to know exactly what b contains. But if we cannot know what
our total evidence is, still less can we know what our total "background
knowledge" is. I dispense with this concept and replace it with the idea
of a historical record of tests in the field of the theory in question. A
test on a predictive consequence of a theory is said to be *hard* if the
consequence is empirically novel, in the sense that no previous test
would have constituted a test of it if it had been formulated then. A test
is *soft* if it can be regarded as a mere repetition of a previous test which
this predictive consequence would have passed. And a test is *medium* if
it is more stringent or searching than the previous tests that this

predictive consequence has passed (or would have passed if it had been formulated then).

I take the total testable content of a theory T, which I denote by $CT(T)$, to be the totality of all its singular predictive implications, that is, consequences of it that say that if such-and-such experimental conditions were realized, then such-and-such an observable effect would follow. In my book I provide rules for what I call a "natural" axiomatization of a theory. And I call an axiom all of whose predicates are theoretical a fundamental assumption, and one, some or all of whose predicates are observational, an auxiliary assumption, of the theory. Now part of the total testable content of T may be entailed just by its auxiliary assumptions alone; and I say that T can gain a corroboration from a favorable test-result only if some or all of its fundamental assumptions were needed for the derivation of the prediction that was tested. If we denote its auxiliary assumptions by A we may represent what I call the corroborable content of T by $CT(T)$ − $CT(A)$. This expression designates all those singular predictive implications of T that are not entailed by its auxiliary assumptions alone.

I say that theory T_1 is better corroborated than theory T_2 if T_1 is unrefuted and no test-result has been more favorable to T_2 than to T_1 while at least one test-result has been more favorable to T_1. A test-result is more favorable to T_1 if it corroborates T_1 without corroborating T_2 or if it refutes T_2 without refuting T_1.

Suppose that T_1 is the best corroborated theory in its field. If it were the sole unrefuted survivor then it would obviously be the best or most preferable theory in its field, given our demand that, to be acceptable, a theory must be possibly true. So suppose that there is another unrefuted theory, say T_2, in its field and that T_1 is better corroborated than T_2. Can we conclude that T_1 is the best theory in its field at the present time, in the sense that it fulfills the optimum aim for science better than all its rivals?

To answer this we need to know something about what tests have been made. After all, it *might* be that T_1 is better corroborated than T_2 because *no* tests have been made on T_2 while just one test, with a favorable outcome, has been made on T_1. So I will introduce the following assumption. If either theory has excess corroborable content over the other, in the sense that it yields predictions in an area where the other theory is silent, then at least one test has been made on this excess content.

Since T_1 is better corroborated than T_2 there is at least one test-result that is more favorable to T_1. This could mean either (i) that the result corroborated T_1 but not T_2 or (ii) that it refuted T_2 but not T_1. However (ii) is here excluded because T_2 is unrefuted. So T_1 must have excess corroborable content over T_2. On the other hand, T_2 cannot have excess corroborable content over T_1. For suppose that it did. Then by the above assumption there would have been at least one test on this excess content; and this test would have resulted in either (i) a corroboration for T_2 but not for T_1, or (ii) a refutation for T_2. But (i) is ruled out by our supposition that T_1 is better corroborated and hence that no test-result has been less favorable to it than to T_2, and (ii) is ruled out by our supposition that T_2 is unrefuted.

We may represent the corroborable content of these two theories by, respectively, $CT(T_1) - CT(A_1)$ and $CT(T_2) - CT(A_2)$. And as we have just seen, the former is larger than the latter. Now the main idea behind the technical elucidations in my book of the notion of one theory being deeper-and-more-unified than a rival is this: the deeper theory has the greater testable content, *not* because it is fitted out with a more powerful set of auxiliary assumptions, but because its *fundamental* assumptions (when married to appropriate auxiliary assumptions) are more fertile in predictive implications than are those of its rival. For this to be the case, it is not sufficient that the total testable content of T_1 is greater than that of T_2 (since that might have been achieved merely by more powerful auxiliary assumptions); but it is sufficient that the *corroborable* content of T_1 is greater than that of T_2; which is what we have here. So in the present case we can say that, while both T_1 and T_2, being unrefuted and possibly true, satisfy our modified version of component (A) of the optimum aim for science, the better corroborated T_1 satisfies component (B1—2) better than T_2. And if it does this it will automatically satisfy (B3—4) better: a deeper theory is bound (in accordance with the anti-trivialization principle) also to be wider, or to have more explanatory and predictive power at the empirical level, although the converse does not hold (a theory may be wider but not deeper than a rival, for instance in virtue of more powerful auxiliary assumptions).

Thus on our assumption that at least one test has been made on any excess corroborable content that either of the two theories may have, we can conclude that the best corroborated theory is, at the present time, the best theory, that is, the theory that best fulfills the optimum aim for science.

Many philosophers assume that if Humean scepticism cannot be repelled, then there can be no scientific rationality; for surely the aim of science is to establish theories, if not as true, then at least as more or less probably true? This assumption plays into the hands of enemies of science, just because scientific theories cannot possibly be established as probably true. There is no valid rebuttal of Humean scepticism. Now it is not rational to set oneself an aim that simply cannot be fulfilled; an aim must be feasible. So the first step, in rescuing the rationality of science, must be to discard this infeasible element from a proposed aim for science. And once we discard it, we can reintroduce all those highly desirable elements that had been driven out, at its behest, during the antidepth war. We can have a richly ambitious aim for science. And with the replacement of that old, impoverishing aim by this one, we restore the idea of the rationality of science. For this aim, though lofty, is sharply defined, and in such a way that we can normally tell which of several competing theories best fulfills it. Thus we can make rational preferences between existing theories, and we can know in advance what a new theory would need to achieve for it to be an advance on the best of the existing ones. The rationality of science is vindicated.

The London School of Economics

NOTES

[1] Bertrand Russell. *The Impact of Science on Society*. London: George Allen and Unwin, 1952, p. 9.
[2] In the Preface to the *Prolegomena* (p. 258 of Vol. *IV* of the Berlin Academy edition of the Collected Works).
[3] Moritz Schlick. *Allgemeine Erkenntnislehre*, 1918 (p. 73 of the Enlglish translation, *General Theory of Knowledge*, by A. E. Blumberg. New York: Springer-Verlag Wien, 1974).
[4] *Wissenschaftslehre* Section 161 (pp. 238 f in Bernard Bolzano. *Theory of Science*, translated by Rolf George. Oxford: Blackwell, 1972).
[5] Rudolf Carnap. *Logical Foundations of Probability*. London: Routlege and Kegan Paul, 1950, pp. 70 f.
[6] Ibid., pp. 116 f.
[7] Jaakko Hintikka. *Logic, Language-Games and Information*. Oxford: Clarendon Press, 1973, Chapters. I and XI, and *Knowledge and the Known*. Dordrecht: D. Reidel, 1974, Chapter 7.
[8] Harold Jeffreys. *Theory of Probability*, second edition. Oxford: Clarendon Press, 1948, pp. 100 f.
[9] Karl R. Popper. *The Logic of Scientific Discovery*. London: Hutchinson, 1959, Appendix *viii.

¹⁰ At the end of Book I, Part IV, Section II, of *A Treatise of Human Nature*.

¹¹ He first published this idea in 1933 (see Appendix *i of *The Logic of Scientific Discovery*).

¹² *Realism and the Aim of Science*. Edited by W. W. Bartley III. London: Hutchinson, 1983, p. 58.

¹³ *Objective Knowledge*. Oxford: Clarendon Press, 1972, p. 18.

¹⁴ A. J. Ayer, in *The Problem of Knowledge*, Harmondsworth: Penguin Books, 1956, said that there are certain unbridgeable logical gaps that we are simply to take 'in our stride' (p. 80). He did not say when a gap becomes *too* wide to be taken in our stride.

¹⁵ P. F. Strawson, in *Introduction to Logical Theory*, London: Methuen, 1952, said that evidence may *conclusively* establish a theory even though it does not entail it (p. 234). He did not spell out what conditions must be satisfied for this important relation of non-deductive proof to hold. On the contrary, he insisted that no such conditions can be specified (p. 248). He adopted a "sealed lips" policy: *there exist* valid non-deductive inferences but we *cannot say* what their nature is.

¹⁶ Published in 1984 by Princeton University Press in the U.S. and by Hutchinson in the U.K.

¹⁷ *The Logic of Scientific Discovery*, Section 83.

¹⁸ "The Maturation of Science", in Imre Lakatos and Alan Musgrave (eds.). *Problems in the Philosophy of Science*. Amsterdam: North-Holland, 1968, p. 136.

¹⁹ *The Logical Foundations of Probability*, p. 575.

²⁰ "'Degree of confirmation' and Inductive Logic", in P. A. Schilpp (ed.). *The Philosophy of Rudolf Carnap*. La Salle: Open Court, 1963, p. 779.

²¹ *The Philosophy of Rudolf Carnap*, p. 988.

RAIMO TUOMELA

SCIENCE, PROTOSCIENCE, AND PSEUDOSCIENCE

I. THE METHOD OF SCIENCE

1. It is often thought that human knowledge must be grounded on some immutable fundamental principles if it is to be reliable and adequate. Thus Kant claimed that human experience is possible only if there are such underlying principles which are *a priori*, viz., known without appeal to experience and also necessary and universal. He called these principles transcendental ones.

Kant's transcendental foundationalism can, however, be put under heavy fire (see, e.g., Rorty, 1979, and Tuomela, 1985). I shall not here present direct criticism against such strong transcendentalism but only refer to an alternative way of looking at these matters. This view builds on an anthropological view of man and claims that all knowledge — philosophical, scientific, mathematical, and common sense knowledge — should not depend on any fixed and immutable prior principles in something like the Kantian sense. Instead it is emphasized that human beings are plastic, constantly changing and evolving natural beings. Human knowledge has no immutable foundation. To be sure, it depends on many things — there is no knowledge without prior assumptions. But such prior knowledge is mutable — it can be questioned and changed, too.

Transcendental assumptions and principles in something like the Kantian sense can be opposed and criticized more directly by arguing them to depend on the so-called Myth of the Given. Generally speaking, this myth involves the ideas that the world, our knowledge of it, and our language are immutably given to us. Thus the world is assumed to be *a priori* categorially given, and our knowledge and language about it are taken to be somehow "logically" or "necessarily" connected to it. (See Sellars, 1963, and Tuomela, 1985, for relevant discussion.) On the other hand, the rejection of the Myth of the Given entails that there can be no knowledge without prior knowledge. Accordingly, also a kind of transcendental knowledge, viz. knowledge of knowledge of objects or "metaknowledge", is needed, although not in the Kantian sense.

83

J. C. Pitt and M. Pera (eds.), Rational Changes in Science, 83—101.
© 1987 *by D. Reidel Publishing Company.*

Not surprisingly, the rejection of the Myth of the Given is compatible with a naturalistic-anthropological view of man as an evolving part of nature and society. Such rejection also fits well with a kind of "internal" scientific realism, as argued in Tuomela, 1985. Internal scientific realism — as opposed to "metaphysical" realism — emphasizes that knowledge and truth are necessarily viewpoint-dependent notions. Thus truth is an epistemic notion. (The present remarks are relevant for our discussion of the method of science, because we will claim below that it presupposes the tenability of realism.)

My general argument in this paper is that the scientific method is a most important method for cognizing the world. Indeed, it is so important that it can be regarded as a criterion of what there is and isn't in the world. Or this is at least what the scientific worldview as well as scientific realism claim — *scientia mensura*, it is said. The central argument for this view is that the method of science is simply the best (most reliable, valid, and best-explaining, etc.) method for gathering knowledge about the world. But even if that were so, would this not entail giving the method of science a priviledged status in a foundational sense relying on the Myth of the Given? The answer is no, because the method of science is not immutably given to us but is rather a collection of plastic principles. Let me emphasize already here that it is characteristic of science that it is capable of correcting itself. That is, science is plastic and progressive in the sense that it can correct both its method and its results (the products of that method). In this sense science can be argued (and will be below) to differ strongly from such doctrines as magic, religion, and pseudoscience, which are not self-corrective.

In this connection two short remarks may be inserted before going on. First, what is important in science is its progressive method rather than its substantive content at any time. Secondly, the adoption of the scientific world-view and scientific realism does not entail scientism in any ideological sense and is thus compatible with e.g. "soft" and "green" values.

2. Let us now take a somewhat closer look at science. We shall start with the general features of the scientific method. How can we distinguish between what is science and what is not? This problem — the so-called demarcation problem of science — has been widely discussed in philosophy. We can attempt to give a solution by, e.g., trying to find

some standards for scientific activity and the scientific method such that these standards define what it is to be scientific. If we succeed in this we can build out of scientific elements a worldview which excludes pseudoscientific ingredients (such as voodoo and astrology). Pseudo-science and magic are curses of our time, curses that are supported by irresponsible authorities, by fear of various kinds of catastrophies and by human gullibility. We shall return to pseudoscience later.

Science is an extremely complex phenomenon. We shall explicate it further later on, and the following should suffice here. When we speak of science we can refer to the science institution (organized scientific communities), to the research process, to the scientific method or to scientific knowledge. Scientific communities form the part of society which produces knowledge through research work. Research in its turn can be looked upon as social activity which aims at the systematic and organized pursuit of new knowledge and which follows scientific research methods. Research methods are the means of production of scientific knowledge, accepted by the scientific community. Scientific knowledge, finally, consists of the products of research which have been obtained with the help of these means.

The general characterizations of the above paragraph are perhaps too circular to be very satisfactory for the understanding of the nature of science. For this reason it is desirable to attempt to characterize scientific research activity by features that do not directly draw on scientificalness. Several such features have been described in professional literature — these characterize above all the scientific method. These partly overlapping criteria or characteristics include objectivity, criticalness, autonomy, and progress. The list could be continued and we will continue it later. The above ones may nevertheless be the most central ones and they all (except perhaps autonomy) serve to characterize the rationality (at least means-end rationality) of science. I shall briefly explain them, without however giving detailed justifications.

The criterion of the objectivity of science contains, first of all, the requirement of the objectivity of the domain of study: science examines real things, be they stones, animals, electrons or historical documents. Secondly, science is objective in the sense of intersubjectivity. The agents of scientific inquiry are scientific communities (with their members) and the motivational background of inquiry is formed by the we-attitudes (we-intentions, -wants, and -beliefs) of its members, not the largely idiosyncratic wishes and conceptions of the individual investigators which enter into the process (cf. Tuomela, 1984, 1985). Scientific research process must be public throughout, or at least public in principle. This feature also includes (at least in principle) the requirement of repeatability: e.g. the results of scientific experiments must be reproducible. (The exact content of required repeatability may be debated about but some relevant — perhaps subject matter-dependent — repeatability should be insisted on.)

The critical nature of science incorporates, not so much freedom from all presuppositions (for this is impossible, cf. the Myth of the Given) but a critical and skeptical attitude towards them. In science nothing is immune to criticism — presuppositions, concepts, theories and hypotheses, theoretical inferences, experimental set-ups and the performances of experiments, conclusions drawn from data, etc. are all subject to examination. Of course all criticism and doubt is also based on its own foundations, but these can be varied. The old simile of the scientific enterprise (philosophy included) as a ship which is rebuilt at sea plank by plank, is still appropriate.

There is one outstanding feature associated with the critical nature of science which is especially central, viz. testability. By testability I mean, more particularly, the empirical testability of scientific theories (the nature of which depends on the field of inquiry). The scientific method is liberal in the extreme with respect to theory formation. It is reasonable to allow the presentation of bold and even unlikely ideas. It is important not to suppress scientific creativity — rather, all flowers are allowed to blossom. It is a separate matter altogether that theories and hypotheses are to be testable (and falsifiable), and this requirement must be strictly imposed. Testability itself need be but indirect — i.e. based on auxiliary hypotheses. But the stricter the tests allowed by a theory, the better. If science does not fulfill the requirement of testability it does not reproduce and develop but stiffens and turns into pseudoscience.

Criticalness and testability are closely followed by self-correctiveness — a central feature of the scientific method not possessed by any other method for gathering knowledge about the world. Critical scientific discussion which also reaches the test results of scientific hypotheses, can be expected to lead to a process in which inquiry corrects its own mistakes — be these mistakes in erroneous data or false theories, or even in errors in the scientific method itself.

The self-correcting nature of science was defended with great vigour by C. S. Peirce already in the 19th century (cf. Rescher (1978)). More particularly, he thought that science can eliminate errors. Thus it can be shown for some central scientific tests at least that, when the observations accumulate, an erroneous hypothesis is very likely to be rejected. The same can be said of scientific discussion and debate at large. Unfortunately the matter has not been very thoroughly studied in its general form in the literature (cf. Laudan (1973)). Thus we still lack a unified and precise theory of self-correction, despite the fact that the matter has been examined rather precisely in the case of inductive methods, and despite the fact that cybernetics and theories of learning offer various kinds of feedback mechanisms on which a theory could be based. Since self-correction can apply both to truth and the methods of obtaining truths, such a unified theory must deal with both aspects (cf. Rescher (1977)).

Self-correctiveness is a very clear rationality feature, for it is of course rational to be able to correct one's errors on one's way to best-explaining and true theories of the world. We can accordingly say that a method for gathering knowledge is the more rational the more conducive to truth it is.

One feature associated with the self-correcting nature of science seems nevertheless rather obvious: if science cannot correct its own results and methods, nothing can. It cannot be done by God, the King, the Party or the oracle of Delphi. In this respect science is in fact a self-sufficient institution which does not allow external checks of validity; there are no extra-scientific criteria of correctness (a fact which does not rule out several kinds of connections, factual and other, between institutionalized science and the rest of the society). Indeed, science is self-corrective partly because it is autonomous.

There are plenty of sad examples of the intervention of e.g. political and religious powers in the internal affairs of science. We can, I submit, nevertheless accept the principle of the autonomy of science in the sense that science is or should be autonomous in the choice of the criteria of truth and correctness, as well as in the application of these criteria in scientific inquiry. On the other hand we can see that science can recover from at least minor violations of its autonomy — as we can judge from history (cf. e.g. the Lysenko-affair).

Progressivity was also mentioned as one of the characteristics of the scientific method. It is not possible here to go into this broad matter at all. I have argued in Tuomela (1985), Chapter 9, that progressive scientific change is possible if the scientific method is applied rationally. Thus given that scientists are in certain specific ways rational (e.g. in replacing their theories with new, better-explaining ones) science will indeed grow towards increasingly better-explaining and more truth-like theories about the world.

II. SCIENCE AND PRESCIENCE

1. In what follows we shall take a still somewhat more detailed look at science, one which is not confined to the scientific method. Relying in part on Bunge's (1983) characterization we can look upon science as a cognitive field. We shall say that a cognitive field is a field of human activity in which the aim is to obtain information of a particular area and, in one way or other, to make use of such information. Science is one such cognitive field and religion another, to mention a couple of examples.

A cognitive field, K, can be regarded as an ordered tentuple such that $K = \langle A, Y, F, E, D, S, P, T, G, M \rangle$, where the components of K (understood to be suitable sets, if needed) have the following content:

(1) A represents a scientific community;
(2) Y represents the host society of A;
(3) F refers to the scientific community A's general (philosophical) views or worldview;
(4) E refers to the exact (logical and mathematical) thought equipment employed by A;
(5) D refers to the domain of inquiry of A;
(6) S refers to the specific knowledge and background assumptions of A obtained from other fields;
(7) P refers to the set of problems of A;
(8) T refers to the specific pool of information obtained by A through its action;
(9) G refers to the relevant aims of A;
(10) M refers to the methods employed by A.

As such the component's of K make no reference to either actions or the practical inferences which justify actions. However, they come into the picture in an indirect way, for the community A partly gathers its pool of information about the world via them. This means that A engages in suitable research activity and relies on practical inferences which in fact contain elements from all components of K.

We can now define a particular science such as biochemistry or psychology as a cognitive field K which satisfies at least the following conditions:

(i) All elements in K can undergo some changes; the crucial matter is that (3)—(10) may be altered as a result of research results obtained in neighboring fields (cf. E and S in particular).

(ii) The members of A are rational and well-enough trained to be able to perform relevant practical inferences and to act accordingly (insofar as these inferences and actions are otherwise possible).

(iii) The society Y, or the societies in which the said particular science is practiced, offers the scientific community A autonomy of inquiry and the resources required by its scientific activities.

(iv) The general philosophical views F contain (a) an ontological view of relevant objectively existing real things which can change and act as causal agents, (b) an adequate view of the scientific method and (c) a view of science as organized activity which aims at least in part at factually truthlike descriptions and explanations, as well as (d) ethical rules for conducting research, especially the ethos of the free search for truth, depth, and systemicity.

(v) The exact thought equipment E consists of up-to-date logical and mathematical theories which can be used, e.g., to sharpen theory formation and to process data.

(vi) The object domain D consists of real objects, past, present and future (cf. iv) (a)).

(vii) Specific background knowledge S consists of up-to-date and sufficiently well-confirmed data, hypotheses and theories which have been obtained from the neighboring fields relevant for K.

(viii) The set of problems P comprises, at least primarily, cognitive problems about the domain D and other components of K.

(ix) The specific pool of information T consists of up-to-date, testworthy and testable (and in part tested and confirmed) theories, hypotheses and data which are compatible with F, as well as of special information previously incorporated into K.

(x) The goals and aims G have to do, first and foremost, with the search for and application of the laws and theories about the domain D, with the systematization of the information about D, with the gener-

alization of this information into theories, as well as with the improvement of the methods M.

(xi) M consists of appropriate scientific methods which are subject
to criticism, test, correction, and justification.

(xii) K is connected with a wider cognitive domain K' whose investigators are similarly capable of scientific inference, action and discussion
as the members of A are, and whose members have a similar supporting
society (or organization of such societies), and for which it also holds
that (a) the components F, E, S, T, G, M of K are in part identical with
the corresponding components of K' (i.e. they have a set-theoretically
non-empty intersection) and (b) the domain D' of K' includes D or else
every element of D is a component of some system of D'.

The above characterization, which to a great extent relies on Bunge's
(1983) proposal, contains plenty of elements which cannot be subjected
to further scrutiny here. Nevertheless we can say, quite generally, that it
fits well with the internal realism presented in Tuomela (1985). We can
see this by examining the conditions (i)—(xii), one at a time.

Science, we have observed, is creative social activity which contains
no a priori given or privileged elements. This supports condition (i).
Conditions (ii) and (iii) can be held true even on the basis of what was
said above. These two conditions are nontrivial — not every society is
capable of creating and supporting a scientific community (cf.
theocratic societies). It is equally obvious that science presupposes
philosophical background assumptions (cf. the Myth of the Given).
Whether all assumptions built into (iv) (especially (c)) are required is
perhaps controversial — at least for an instrumentalist. In any case
scientific realism satisfies condition (iv) full well. But as I have shown in
Tuomela (1985), Chapters 4 and 5, (iv) can be supported also without
direct reference to realism. But what is more, it can be plausibly argued
that the method of science simply and bluntly is realism-dependent (see
especially Boyd, 1981, on this). Thus, for instance, science is dependent
on inference, in the right circumstances, to best explanations involving
reference to unobservables, it may be argued. Condition (v) needs no
deep justification. It can hardly be denied that muddled though is far
from scientific thinking. (This condition itself is insufficient to guarantee
clear thinking, nor can it without help from the other conditions
safeguard from quasi-precision.)

Condition (vi) can be grounded on the requirement of objectivity
(cf. above) and on the conceptual and philosophical muddles associated

with other than the ontology of real objects (cf. Tuomela, 1985, Chapter 7).

Conditions (vii)—(xi) are standard assumptions which can be accepted without further ado — they harbor no special controversies. Naturally these conditions in particular satisfy the general characteristics of the scientific method enumerated above (objectivity, criticalness, testability, self-correctiveness, autonomy and progress). Condition (xii) points to the systemic character of science — it must dovetail with some wider cognitive setting. This requirement alone does not rule out e.g. scientific revolutions.

Scientific activity is both end-rational and means-rational when pursued according to (i)—(xii). The gaining of scientific knowledge enhances man's chances for survival — that is a consequence of the successful pursuit of truth (this expresses end-rationality) according to the methods of science (means-rationality). To what extent man is capable of realizing those chances depends on the success of his practical activities and their rationality (practical rationality).

Our above characterization of science may seem too strict. If we consider the social sciences and the humanities we can see that at least conditions (v), (vii), (ix), and (xii) typically are difficult to satisfy. Still we may regard these conditions as acceptable ideals in science. We may accordingly want to speak of fields of research which are scientific in a weaker sense than above. Let us briefly consider this matter.

2. Above we defined the notion of cognitive field as our starting point and went on to characterize science on the basis of it. But we can indeed define interesting intermediate notions (cf. Bunge, 1983). Let us first make a broad distinction between research fields and belief systems and take them to be (at least ideally) mutually exclusive (but perhaps not jointly exhaustive) subfields of the whole family of cognitive fields.

A research field can be viewed as a cognitive field which is based on research and which changes on the basis of research. It is a minimum requirement of a research field that it satisfies, *mutatis mutandis*, conditions (i), (ii), (iii), (viii), (xi), (xii) of the defining characteristics of science. Condition (i) is of course central and so obvious that it needs no further defense here. Conditions (ii) and (iii) should also be obvious once they have been reformulated so as not to make direct reference to science (but only to research). In (xi) we now speak of research methodology or methodics only. Research methodics is a notion which need not involve the method of science in its full sense. What exactly it should be taken to involve is a broad and difficult topic which we cannot discuss here. As to (xii), it must now be reformulated to speak only of research instead of scientific research in the fullest sense.

A belief field (or faith field, if you prefer) is a cognitive field which either does not change at all or changes due to factors other than research (such as economic interest, political or religious pressure, or brute violence). Thus in the prototypic sense a belief system has nothing to do with research and therefore it negates all the defining characteristics of both science and research field. For our present purposes it is not central to give a more detailed analysis of belief fields here.

Among research fields we may count the basic natural sciences, applied sciences, formal sciences (logic, mathematics), technology (including medicine), jurisprudence, the social sciences and the humanities. Among belief fields we include especially religions and political ideologies as well as various pseudodoctrines.

There are various intermediate notions between research field and science. Thus we may call a research field a protoscience or prescience if it is a field on its way to becoming a science. More exactly, we may take a protoscience to be a research field satisfying at least conditions (i), (ii), (iii), (iv), (vi), (viii), (x), (xi), and (xii). In addition, there should be at least some reason to think that a protoscience can develop so as to ultimately satisfy the rest of the defining characteristics of science. On the whole, the social sciences can be regarded as protosciences in the defined sense. It should be emphasized that while protoscience is strictly speaking still non-science it should be clearly distinguished from permanently non-scientific fields such as belief fields (as defined above) and pseudoscience. Let us say here in a preliminary way that a pseudoscience is a cognitive field (often but not always a belief field) which is non-scientific (and typically permanently so) but which its proponents still advocate as science. (There are also pseudo-doctrines which are clearly belief systems and whose proponents do not regard them as scientific but claim the opposite, rather.)

It is appropriate to emphasize in this connection that our classifications represent a kind of ideal types. Note in particular that within a science there may exist non-scientific schools, etc. For some purposes at least it might be better to speak of scientific versus non-scientific attitudes, thinking, and action rather than apply these attributes directly to cognitive fields.

III. PSEUDOSCIENCE

1. The scientific worldview is very rare. My guess is that at least 99% of all currently living human beings have a non-scientific worldview and way of thinking. Most people probably base their lives on religion and/or magic. Not surprisingly, magic and religion are relevant also for pseudoscience, as will soon be seen. But before that let me amuse the reader by mentioning some results a Gallup investigation conducted in the U.S. in 1978 produced. According to it, 57% of all Americans believe in ufos, 54% in angels, 51% in telepathy, 39% in devils, 37% in precognition, 29% in astrology, 24% in clairvoyance, and (only!) 11% in ghosts (cf. Greenwell, 1980). Irrespective of whether this investigation was fully waterproof we can see the general sociological situation in one of the most civilized and educated societies in the whole world. The figures speak for themselves and show both the prevalence of magical and religious thinking as well as its factual connection with pseudoscientific thinking.

To go into more philosophical connections, magic, religion as well as much of pseudoscience deal with an obscure mentalist or "spiritual" ontology (cf. spirits, gods, demons, fairies) and that is true of many pseudosciences as well. We may put this more generally and say that magic, religion and many pseudosciences deal with paranormal entities and phenomena. To say that a phenomenon is paranormal is often understood in the broad sense to mean that it cannot in principle fall within the ontology of science (or, alternatively, at least be explained by science). (A) more specific characterization, defended by Braude (1980), goes as follows. A phenomenon P is paranormal just in case (a) P is inexplicable by current science, (b) P cannot be explained scientifically without major revisions elsewhere in science, and (c) P thwarts our familiar expectations about what sorts of things can happen to the sorts of objects involved in P. But here we need not adopt any specific explication of paranormality.)

Next, the mentioned belief fields typically rely on authoritarian epistemology (cf. God's revelation as a source of certain and immutable knowledge) and dubious aprioristic transcendental principles. Inferences are typically based on dubious a priori principles. Thus in typical magical thinking as well as in much pseudoscientific thinking the so-called 'principle of similarity' is relied on. For instance, in homeopathic magic the magical operations deal with entities which suitably resemble

the ones that one wants to affect. To take an example from homeopathic medicine, still in the last century a preparation made of butter-
cup, the poisonous yellow flower, was thought to cure jaundice: *similia
similibus curantur*! But of course all that is very silly, we now know.
And we also know that it is completely unwarranted that some such
principle of similarity could be a valid law of nature on *a priori*
grounds. If there are relevant similarities in nature it is the task of
science to find them — and in a specific, detailed fashion rather than
in the form of vague over-arching principles. (See Alcock (1981) and
Tuomela (1985) for further discussion on the affinities between pseudoscience and magic.)

2. It is time to go to a more detailed examination of pseudoscience. We
said earlier that pseudoscience is a belief system whose supporters,
however, incorrectly regard it (knowingly or not) as a science or a
branch of science. Can pseudoscience and pseudoscientific thinking
(and action) be characterized in more detail? If we take such pseudosciences as astrology, the theory of biorhythms, suitable parts of
parapsychology and ufology, homeopathy and faith healing we may
arrive at the following view (cf. Bunge, 1983, Radner and Radner,
1982).
     If we think of pseudoscience as a cognitive field, K, of the kind K =
$\langle$A, S, F, E, D, S, P, T, G, M$\rangle$ as above in Section I, we can see that
pseudoscience differs from science in respect of every element of K
(recall the defining conditions (i)—(xii) of science). I shall below take up
some conspicuous differences. Probably nothing like strict necessary
and sufficient conditions of pseudoscientific thinking can be given. We
must instead deal with a kind of ideal type here. The following list
characterizes it:

(1) A pseudoscience typically or often relies on (a) a very obscure and
ill-defined ontology (consisting of, e.g., unembodied spirits — cf.
witchcraft, parapsychology) and on (b) an epistemology which accepts
epistemic justification on the basis of authority or is based on some
alleged paranormal epistemic abilities of an elite such as priests, and on
(c) a dogmatic attitude and ethos to defend a doctrine even by non-
scientific means as contrasted with scientific search for truth (cf.
creationism, Lysenkoism). Criticism is not welcomed by pseudoscien-

tists. (Recall, on the other hand, the properties (iv), (vi), (x), and (xi) characteristic of science.)

(2) Pseudoscientific thinking often shuns conceptually and logico-mathematically exact thinking — cf. creationism, witchcraft, psycho-therapy. (Recall (v) of the properties of science.)

(3) The hypotheses and theories — to the extent there are any — of a pseudoscience are generally either impossible to test or at least they are badly supported, both empirically and theoretically. This feature is indeed characteristic of all pseudoscience. Pseudoscientists often (incorrectly) try to compensate for qualitatively poor evidence by piling up great quantities of such evidence. Parapsychology provides a typical example of this. (Cf. (vii) and (ix) in the case of science.)

(4) The hypotheses and theories of pseudoscience are not changed due to confrontation with empirical and other evidence. They often flatly contradict well-confirmed scientific hypotheses, but this fact does not usually have much effect on pseudoscientists, who "know better". On the whole very little change takes place in pseudoscience and such change is typically due to factors not related to research. (Recall parapsychology, Lysenkoism, flatearthism, etc. and contrast with properties (i), (vii), and (ix) of science.)

(5) Pseudoscience involves anachronistic thinking, viz. thinking going back to old, refuted theories and assumptions; cf. e.g. creationism, flatearthism, and confront this with features (v), (vii), and (ix) charac-terizing science. For instance, creationists argue against evolution theory by claiming that mutations always involve something harmful to the organism — but this is a false and refuted claim.

(6) Pseudoscience often appeals to myths (cf. von Däniken on ancient astronauts, creationism) and unfounded mysteries (cf. the Bermuda Triangle, ufos, poltergeists). (Recall characteristics (iv), (v), (vii), (viii), (ix), and (x) of science.)

(7) The problematics of a pseudoscience often consist in practical rather than cognitive problems (cf. such problems as how to bring about a certain effect, for instance, how to feel better or how to cure a certain illness or disorder). Thus indeed pseudoscience often contains aspects of pseudotechnology. (Cf. features (viii) and (x) of science.)

(8) The methods of pseudoscience are unscientific especially in the sense that they are not self-corrective and checkable by alternative (especially scientific) methods nor are they based on well-confirmed general theories. This applies to all typical pseudoscientific doctrines. (Cf. characteristic (xi) of science.)

(9) Pseudoscience is generally a doctrine or body of doctrines isolated and distinct from the science of its time. (Cf. creationism, Lysenkoism, flatearthism and feature (xii) of science.)

Our above characterization of pseudoscience shows that criteria (i), (iv)-(xii) of science do not apply to pseudoscience but rather their opposites do. As to features (ii) and (iii), the members of a pseudoscientific community can be called believers, and typically they lack the education and skills of proper scientists. The societies which support pseudoscience do it mostly through tolerating it rather than actively providing it with resources — unless perhaps big money is at stake or the political or religious system of that society requires it (cf. Lysenkoism).

Such examples of pseudoscience as the theory of biorhythms, astrology, dianetics, creationism, faith healing may seem too obvious examples of pseudoscience for academic readers. A more exciting case is provided by parapsychology, for it is, indeed, a very controversial field. Let us have a brief look at it.

3. Parapsychology supposedly studies parapsychological phenomena, viz. phenomena which are both psychological and paranormal (recall our above characterization). These are commonly classified as phenomena of telepathy, clairvoyance, precognition, and psychokinesis. The first three of these are jointly called ESP-phenomena ('ESP' standing for 'extrasensory perception'). All parapsychological phenomena are jointly called psi-phenomena. Now it is a very peculiar feature about parapsychological research that it is almost exclusively concerned with showing the existence of psi-phenomena. As far as I know there is no proper science the existence of whose subject matter would be similarly under doubt.

Is there good evidence for parapsychological phenomena, then? If you ask the man in the street (or even a typical academic parapsychologist) he is likely to say that, for instance, telepathic phenomena exist (recall the Gallup result referred to earlier). However, it seems reasonable to oppose him. Indeed, I would like to claim, contrary to the opinion of many parapsychologists, that there still are no repeatable experiments for showing the existence of parapsychological phenomena. That is, there are no recipies for creating conditions under which such phenomena will always or typically take place. And this of course makes the alleged existence of paranormal phenomena rather dubious.

In fact, the situation is worse. For claims about the existence of paranormal phenomena in scientific parapsychology are usually based merely on the existence of statistically significant deviations from either randomness or 'normality', as the case may be; and this is of course a very weak criterion of existence. And yet there are no repeatable experiments even for showing such statistically significant deviations in people's relevant abilities to behave. Thus there are no repeatable experiments indicating the existence of paranormal abilities, still less is there evidence for invoking telepathy, clairvoyance, precognition, and psychokinesis as best-explaining factors for such alleged deviating behaviors.

I have above made strict claims. How can they be substantiated? If you look at parapsychological literature you get the impression that the existence of parapsychological phenomena has been clearly proved (see e.g. Wolman, 1977, Grattan-Guinness, 1982, Eysenck and Sargent, 1982). However, critical investigation has shown that such claims are not well taken at all (see e.g. Alcock, 1981, Marks and Kammann, 1980, Johnson, 1980, Frazier, 1981, Randi, 1982, Hansel, 1980, Gardner, 1982, Abell and Singer, 1981, Hyman, 1982). Perhaps the most important reason for saying so is simply that the experiments have been badly done, even if ostensibly positive evidence for psi-phenomena seems to have been obtained. We cannot here go into any detailed criticisms of experimental faults, cheating and other factors claimed to mar parapsychological experiments. Anyhow, the general conclusion inferrable from the mentioned critical literature is that there is nothing in the experimental results to convince at least a skeptical (but open-minded) observer.

I should here also emphasize the often mentioned feature that there are no proper theories in parapsychology, not even 'models' in the sense there are in otherwise comparable fields in experimental psychology. But there cannot be very 'hard' data without theories. Unless alleged experimental facts can be backed by theories they are often not well confirmed but rather soft.

What disturbs me much as a philosopher is that parapsychological phenomena, if they existed, would contradict (or at least seem to contradict) well-established scientific theories. Only telepathy at very short distances would seem to be compatible with fundamental physical theories. In other words, the existence of parapsychological phenomena would require a drastic revision in our fundamental scientific theories. And that is a consideration which certainly must affect one's evaluation of the alleged parapsychological facts. (Parapsychologists might want to say here that parapsychology deals with unembodied spirits which are not bound by the laws of physics — but that would be an irresponsible response unless clear conceptual sense is made of such paranormal phenomena and until their existence is proved in a strong sense satisfying a skeptic.)

Let us finally compare parapsychology with the properties of pseudoscience as stated earlier. Referring to our earlier numbering, much of parapsychological research satisfies the central criteria of pseudoscience. To wit, parapsychology relies on an ill-defined ontology (feature (1)) and typically shuns exact thinking (2)). The hypotheses and theories of parapsychology certainly are in bad shape (criterion(3)). Extremely little progress has taken place in parapsychology on the whole and, furthermore, parapsychology conflicts with established science (4)). Parapsychology is also poor in its research problems, being mainly concerned with establishing the existence of its subject matter and having practically no theories to create proper research problems (feature (7)). While in parts of parapsychology there are attempts to use the methods of science there are also unscientific areas; and in any case parapsychological research can at best qualify as prescientific because of its poor theoretical foundations (cf. (8)). Furthermore, as emphasized, parapsychology is largely an isolated research area (cf. (9)).

How about criteria (5) and (6)? Does parapsychology involve anachronistic thinking and rely on myths? Perhaps we must answer negatively in the case of much of experimental parapsychology, but there are anyway more exotic fields like the study of reincarnation, out-of-body experiences, and poltergeists (and other ghostly phenomena) in which many kinds of mythical ideas have been relied on (see Alcock, 1981). Indeed, parapsychology is both historically and systematically connected with magic and religion. Its obscure ontology certainly has much to do with this (cf. ghosts, spirits, demons). Also some of the nonscientific thinking to be found in parapsychology relates to this. Let

me mention a couple of examples relevant to experimental parapsychology as well. First, if it is taken for granted (as it is in some psi-
circles) that dreams regularly (or often) serve to predict the future such
thinking relies on backward causation (maybe operant conditioning is
psychologically involved here!). Or think of homeopathic magic (e.g.
voodoo). Accepting the principle of homeopathic influencing clearly
entails the acceptance of psychokinesis.

On balance, there are then good grounds for regarding much of
parapsychology as pseudoscience. However, I would still not quite join
Alcock and many others in regarding all of parapsychology as pseudoscientific — it is partly prescientific, I would rather say. But I think it is
reasonable to require strict demonstration of the existence of psi-
phenomena in view of (1) the rather incredible history of cheating in
parapsychology and (2) the incompatibility of parapsychology with
well-established theories and laws of science. And presently we do not
have such a waterproof demonstration.

4. One could write a lengthy book on different pseudosciences. Space
does not allow me to attempt to demonstrate the pseudoscientificality
of different disciplines. Let me merely mention some examples of
doctrines which either have been convincingly shown to be pseudosciences or which at least are prime candidates. The following short list
speaks for itself: astrology, parapsychology (in part), ufology (in part),
the three-stage theory of biorythms, iridology, Lysenkoan biology,
creationism, (some) empirical applications of catastrophe theory,
anthroposophic agriculture, homeopathic medicine, and some forms of
psychotherapy.

Pseudoscientific thinking is widespread. It flourishes all over, even
within scientific communities. (Doctrinal edifices are seldom entirely
monolithic — science can contain pseudoscientific islands and pseudosciences scientific islands.) Pseudoscience is much more popular and
profitable than science. It brings about troubles to scientific policy
makers at least when they do not have a sufficient scientific background. Pseudoscience is more of a headache for laymen than for
scientists or scholars, for the latter, or scientific communities in general,
easily distinguish pseudoscience from science on the grounds that it
does not apply the scientific method.

A greater difficulty for science is to distinguish from one another
a budding, promising, and possibly unorthodox protoscience, and

100   RAIMO TUOMELA

enticing unorthodox research which eventually turns out to be an unproductive hybrid and possibly a pseudoscience (cf. the so-far unsettled fate of sociobiology). Both may instantiate some sort of application of the scientific method — the rejection of Wegener's theory of continental drifts for half a century is an example of groundless rejection of a protoscience. We have here a case where a scientific community has dogmatically committed itself to normal science.

We can note that the problems for a layman are magnified. Many times he ought to be able to distinguish between pseudoscientific and scientific inquiry, and within science between unfruitful and really significant research. This problem is especially pressing in areas where science still is at an infant stage or at least has not reached grounded applications; and this situation, as we know, is not uncommon. As examples we could mention problems of education, of nourishment, as well as the multifarious difficulties having to do with health care and medication.

It is not possible here to discuss in more detail how to distinguish between pseudoscience, prescience, and science in actual practice. There are clearly lots of borderline cases and drawing borders is not easy. However, as we have argued and elucidated in this paper, it is at least in principle possible to distinguish between scientific, prescientific, and pseudoscientific thinking and to regard these distinctions as both theoretically and practically significant.

                               REFERENCES

Abell, G. and Singer, B. (1981). *Science and the Paranormal*. London: Junction Books.
Alcock, J. (1981). *Parapsychology: Science or Magic?* Oxford: Pergamon Press.
Boyd, R. (1981). "Scientific Realism and Naturalistic Epistemology". In Asquith, P. and Giere, R. (eds.). *PSA 1980*, Vol. II. East Lansing: Philosophy of Science Association. Pp. 613–662.
Braude, S. (1980). *ESP and Psychokinesis*. Philadelphia: Temple University Press.
Bunge, M. (1983). *Treatise on Basic Philosophy*, Vol 6. Dordrecht: D. Reidel.
Eysenck, H. and Sargent, C. (1982). *Explaining the Unexplained*. Willemstad: Multimedia Publications.
Frazier, K. (ed.) (1981). *Paranormal Borderlands of Science*. Buffalo: Prometheus Books.
Gardner, M. (1982). *Science: Good, Bad, Bogus*. Buffalo: Prometheus Books.
Grattan-Guinness, I. (ed.) (1982). *Psychical Research*. Wellingborough: the Aquarian Press.
Greenwell, J. (1980). "Academia and the Occult". *The Skeptical Inquirer* 5, 39–45.

Hansel, C. (1980). *ESP and Parapsychology*. Buffalo: Prometheus Books.

Hyman, R. (1982). "Does the Ganzfeld Experiment Answer the Critics' Objections". In *Program and Presented Papers*, Vol. I of the Centenary Jubilee Conference of the British Society for Psychical Research.

Johnson, M. (1980). *Parapsykologi*. Uddevalla: Zindermans.

Laudan, L. (1973). "Peirce and the Trivialization of the Self-Correcting Thesis". In Giere, R. and Westfall, R. (eds.). *Foundations of Scientific Method: The 19th Century*. Bloomington: Indiana University Press. Pp. 275—306.

Marks, D. and Kammann, R. (1980). *The Psychology of the Psychic*. Buffalo: Prometheus Books.

Radner and Radner (1982). *Science and Unreason*. Belmont: Wadsworth.

Randi, J. (1982). *Flim-Flam*. Buffalo: Prometheus Books.

Rescher, N. (1977). *Methodological Pragmatism*. New York: New York University Press.

Rescher, N. (1978). *Peirce's Philosophy of Science*. Notre Dame: University of Notre Dame Press.

Rorty, R. (1979). *Philosophy and the Mirror of Nature*. Princeton: Princeton University Press.

Sellars, W. (1963). *Science, Perception, and Reality*. London: Routledge and Kegan Paul.

Tuomela, R. (1984). *A Theory of Social Action*. Synthese Library. Dordrecht: D. Reidel.

Tuomela, R. (1985). *Science, Action, and Reality*. Dordrecht: D. Reidel.

Wolman, B. (ed.) (1977). *Handbook of Parapsychology*. New York: Van Nostrand.

THOMAS NICKLES

METHODOLOGY, HEURISTICS, AND RATIONALITY[1]

> These rules of old discover'd, not devis'd,
> Are Nature still, but nature methodiz'd.
> — Alexander Pope

> What I call 'methodology' should not be taken for an empirical science . . . . Methodological rules are . . . conventions . . . the rules of the game of empirical science. — Karl Popper

> Each chief step in science has been a lesson in logic.
> — Charles Peirce[2]

I. INTRODUCTION

The discussion of method has a long history. The term 'method' is Greek (from *meta* + *odos*, meta-way, "following after," suggesting the idea of order), and already in Plato's *Phaedrus* (265D to 277C) Socrates speaks of method in relation to *téchnē* or art. Socrates refers to the method of Hippocrates, presumably because Hippocrates's method or art of medicine (1) was useful in solving practical problems of life, (2) guided practice with understanding and was thus superior to both (3) those medical skills which rested on mechanical, routine procedures and (4) those inefficient procedures which succeeded by chance or accident. Procedures of kinds (3) and (4) are both "blind." Indeed, Socrates says that any artist attempting to proceed without method is like a blind man (Gilbert, 1960, pp. 3, 4, 39—41).

Although there have been many shifts in the meaning of 'method' through the centuries, I shall take these early ideas — plus the 16th and 17th century insistence on efficient, systematic ordering — as fundamental. In particular, I shall treat method and methodology as concerned primarily with inquiry and the theory of inquiry, respectively. Methodology is deeply concerned with the process of investigation, with doing. Hence, I am in basic agreement with Clifford Hooker's characterizations of method and methodology:

103

J. C. Pitt and M. Pera (eds.), Rational Changes in Science, 103—132.
© 1987 *by D. Reidel Publishing Company.*

Method describes a sequence of actions which constitute the most efficient strategy to achieve a given goal; *Methodology* describes the theory of such sequences. . . .

According to the dominant English-speaking philosophic and scientific tradition method is determined by theory of science and more particularly by epistemology. According to another tradition, method defines the scope of epistemology and determines the form of the theory of science. With the former view we associate the empiricist mainstream of this century, with the latter the pragmatists (e.g. Dewey) and Popper. Both approaches can accept the dictum that it is method which distinguishes the practice of science from, say, myth. (1977, p. 3.)

I am also in agreement with many ideas of Nicholas Rescher's methodological pragmatism (1977, 1978).[3] However, I think of methods as detailed, content-specific (and therefore usually quite local) problem-solving procedures, whereas Rescher thinks of method in a more global sense as concerned with the philosophical presuppositions and procedural strategies (such as induction or hypothetical-deduction) that all sciences, and perhaps all inquiries, possess in common. One could follow Rescher (p. 2) in employing the term *metamethodology* for the comparative study of competing methodologies, but I shall simply lump everything above method into 'methodology'.

I shall spell out this minority conception of methodology in my own terms, with three departures from Hooker's formulation. (1) Since there are bad methods as well as good ones, we cannot describe method as *the most efficient* strategy. It is precisely the job of methodology to determine which method is most efficient for achieving a given goal. (2) As Herbert Simon (1957) insists, human rationality is bounded; so we usually must be content with methods which are *efficient enough* (sufficiently efficient) — with "satisficing" rather than "maximizing" efficiency. (3) I give Popper's methodology mixed reviews. Popper is closer to the dominant British empiricist tradition than Hooker allows, and he retains its aprioristic methodological tone. Moreover, Popper has no methodology (or very little) in Socrates's sense, not to mention Peirce and Dewey. Rather, he conceives of method as a mechanical procedure and concludes that there is no scientific method, properly speaking. Since scientific discoveries are not made algorithmically, discoveries must be lucky guesses, the products of chance and accident. Popper misses Socrates's and Peirce's point that method falls between these two extremes.

There is another way in which the account of method and methodology which I defend locates them between two extremum positions. At one extreme is the view captured in Pope's lines: method is the "action" form of a strongly realistic and epistemically optimistic theory of nature. Thus, method is a reflection of empirical (and perhaps meta-physical) reality — a reality that is knowable and, to a significant extent, already known. Pope's poetry reflects Newtonian epistemological optimism: "God said, *Let Newton be*! and all was light." At the other extreme is· the aprioristic view that method is essentially logic plus conventional rules and hence devoid of empirical and metaphysical content, method and methodology assert or presuppose nothing about the world: Method is neutral among competing theories and is in no way discipline- or theory- or problem-specific.

As before, Peirce and Dewey fall between these extremes. Popper, who rejects the "naturalistic theory" that methodology is an empirical science, holds a liberal version of the second extremum position. Popper *claims* that his epistemology is largely derivative from his scientific methodology, but that may be doubted (see §III below). Some positivists, e.g., Carnap (1934, 1935), held a more reductivistic, epistemological-foundational view of methodology as derivative from epistemology, with epistemology in turn a branch of logic. Hence, methodology of science reduces to logic of science. Other positivists were more naturalistic.

A central question of methodology is the source or basis of the normative force of methodological rules and recommendations. Again, I stand with the pragmatists in seeing methodology as prior to academic epistemology rather than as "applied epistemology" — the epistemology of X, where 'science' is now substituted for the variable 'X'. Major changes in methodology are the result of substantive changes in science rather than of major changes in philosophy, as my Peirce motto suggests (cf. Laudan, 1977). Roughly speaking, I defend a view of method more naturalistic than the logical-conventionalist view (which fails to explain where logic comes from, unless it is mere linguistic convention — whence it fails to explain how method can direct inquiry and why it should be at all successful in doing so) but less naturalistic than Pope's Newtonian standpoint.

The dimension missing from both these extremum positions is the pragmatic, "how to do it" dimension, which is bound up with human psychology and sociology and with limited human resources. Limited intellectual and supportive resources make economy of research a central topic. Also, science as we have it is a human, social activity which seeks to achieve humanly formulated goals by humanly possible means. Against Pope, the order of knowing (*ordo cognoscendi*) is not identical with the order of being (*ordo essendi*). But neither is the order of doing (*ordo faciendi*) identical with the order of knowing, let alone with the logical order. Hence, methodology, although related to theory of nature, to particular scientific theories, to epistemology, and to logic — cannot be identified with, or reduced to, any of them. Nor, in my opinion, can method be identified with a set of rules. For scientists can learn what to do next and how to do it otherwise than by learning rules. As Dewey (1938), Wittgenstein (1964), Polanyi (1958), Kuhn (1962), and Toulmin (1972) each have stressed in their own way, we can learn from example by doing. Initially, we learn by example to apply standard tools to exemplary problems already solved by previous generations of investigators. Later, we become adept at stretching old solutions into new forms to meet new problems. Even Descartes stressed that the four rules of his *Discourse on Method* were rather vacuous unless their distillation from his scientific practice was studied carefully.

In outline, the connections I see among my topics of methodology, heuristics, and rationality are these. As the pragmatists emphasized, the function of inquiry is to solve problems, that is, to overcome obstacles which prevent us from reaching desired goals. At one level, rationality concerns finding and using the best (or adequate) means of reaching those desired ends — the best method of problem solving. Determining the best way ("method") to get what we want is the task of methodology. Methodology is a practical subject, not one totally lost in high intellectual abstractions. Now 'best' means the most efficient or most effective method, subject to the constraints on the problem. This is where heuristics (in a broad sense) comes in. For heuristics is the study of problem-solving methods and includes a comparative evaluation of the most efficient search and generation methods, given the information at hand. Hence, heuristics constitutes a large part of methodology. Clearly, heuristics does not consist entirely of logical and conventional rules (although the latter have a role in heuristics): we make genuine, empirical discoveries in the field of heuristics — we find out from bitter experience which methods work well and which do not — and the most detailed and useful heuristics tend to be content-specific. We can only find out what works best in practice by examining previous practice (the history of previous efforts) and experimenting with future practice. The most obvious logical-conventional norms may fail utterly in practice (cf. Kuhn, 1962, against Popper).

At another level, rationality concerns the wise choice of goals, given the methods and information at our disposal. In the immortal words of Ralph Barton Perry, there are two ways to solve a problem. You either get what you want or want what you get (Reitman, 1964, p. 308). Hence, there is a firm pragmatic basis for rational criticism of goals.

There are two further points which I can only mention without discussing. First, decisions and decision theory will have an important role in methodology, at several points. Second, one of these points concerns the decision whether to hold out for the most efficient strategy or whether one can, in particular cases, be satisfied with one which is less than optimal. Here I side with Simon (1957) and March and Simon (1958) in saying that "satisficing" often is preferable to "maximizing"; indeed, strict maximizing is rarely humanly possible ("the principle of bounded rationality"). Thus we cannot identify rationality with maximal efficiency. Defining methodology in terms of maximization would make it a very impractical subject indeed. Stated in terms of Peirce's highest methodological rule — "Do not block the road to inquiry" — one important implication is this. Our overarching goal in science, as in life, is not to solve just one problem but to solve as many problems as we can, as rapidly as we can. Hence, a perfectionistic holding out for the *most* efficient choice of method for dealing with certain problems usually will block inquiry; less exacting criteria of success may be, or *must* be, good enough for the purpose at hand. The same point can be made about truth: Do not block the road to inquiry by an obsession with immediate and absolute truth.[4] These maxims do not endorse a cavalier attitude toward efficiency, truth, and precision; but they do run against the foundationist psychology which urges us to get everything absolutely right before proceeding to the next step.

II. SOME HISTORICAL BACKGROUND

In 16th-century Europe, method became a lively topic of discussion as Peter Ramus and others attempted to reform traditional logic and rhetoric. These logicians and rhetoricians placed more emphasis than the Greeks did on efficiency, systematicity, and order. Although method as Ramus, Digby, Temple, *et alia* discussed it had little to do with science as we know it, 16th-century issues and concerns were transformed by Bacon, Galileo, Descartes, Newton and others into an ardent quest for the efficient discovery of new truths. They were interested both in discovery and in justification, as we should say today.

They were equally concerned with the discovery of *new* truths and with the discovery of new *truths*. For lovers of methodological preaching and practice, the 17th century was a wonderful time!

Despite sincere attempts of writers like Peirce, Dewey, and Popper to convince us of "the supremacy of method" and despite the lip service paid to the idea that science is better defined as a method of inquiry than as a body of substantive results, method as a *philosophical* topic has declined since the 17th century and has declined especially rapidly since the mid-19th century. Methodology, as Hooker hints, gradually was reduced to a footnote of epistemology and then, still more narrowly, of formal logic. (Peirce and Dewey, too, identified logic with methodology, but by broadening logic into theory of inquiry — inquiry into inquiry — rather than by narrowing methodology.[5] Compare Popper's identification of epistemology with "logic" of science.) Heuristics in particular was reduced to a dispensable, psychological crutch, of no logical or epistemic significance. Fortunately, there are signs of a revival of interest in methodology in the old sense, in fields such as cognitive psychology and artificial intelligence — and even history and philosophy of science. Now I do not mean that 17th-century methodologies were correct and unobjectionable. Nor, obviously, do I mean that there have been no important methodological advances over the past few centuries. Working scientists have developed a host of conceptual tools, from the calculus and probability theory to modern factor analysis and experimental design. But philosophers interested in science (including Dewey) have had little to do with these developments and, until recently, have taken little professional interest in them and their significance for theory of inquiry.

Epitomizing both the decline and the revival of methodology is the treatment of Descartes by philosophers. Descartes has been historically reconstructed in our own image and then postulated as the founder of "modern" philosophy. Particularly in English-speaking countries, and particularly in the 20th century, Descartes is portrayed as the arch epistemological foundationist — thus making 20th-century philosophy of science seem more continuous with tradition than it is. Many wise scholars have assured us that Descartes attached little importance to empirical data, that his primary concern was to achieve certainty concerning his, God's, and the world's existence, that his aims were theoretical rather than practical, that he conceived scientific method as merely the application of universal reason, and so on. Actually, it is clear from Descartes's own writings that (1) the scientific works and the *Principles of Philosophy* (the scientific portions of which were translated into English for the first time in 1983) are better guides to Descartes's thinking than the *Meditations*; (2) Descartes's foundational tendencies, like Bacon's, were more an attempt to achieve economy of research than simply to combat philosophical skepticism, as the early *Rules for the Direction of Mind* makes clear; (3) among Descartes's chief concerns was the development of heuristic discovery methods (4) in order to make practical progress toward improving human comforts and, especially, human health and longevity and (5) in order to have his new physics and system of the world replace the Aristotelian system in the schools and in scientific work. Recent work has begun to correct the English-language portrayal of Descartes (Sakellariades, 1980, 1982; Clarke, 1982; Gaukroger, 1982; Garber, 1983). It is arguable that Descartes himself saw methodology as prior to epistemology and certainly not as entirely derivative from it. Nonetheless, it must be admitted that Descartes and other 17th-century methodologists too closely linked truth with utility (taking for granted that the true theory was maximally useful) and the order of doing with the order of metaphysical nature, which they optimistically believed to be knowable. Many 17th-century thinkers held near-infallibilist theories of knowledge and inquiry, which everyone today rejects as too strong. Yet, again, I suspect that utility was as important as truth to Cartesian science. After all, Descartes thought he already had achieved understanding of metaphysical reality. *That* was not the job of science (as it is for some of us). Accordingly, the aim of science was to furnish *useful* knowledge. Had Descartes become aware of the fact that the true theory is not necessarily the most useful, Descartes might have diminished the importance of truth rather than that of utility in his methodology.

Ironically, the explosion of writing on philosophy of science in the 20th century has, on the whole, further choked off rather than nourished methodology. This is especially true of the more positivist and formalist philosophies. The promising methodological writings of the American pragmatists — especially Peirce and Dewey — early in the century soon disappeared from view as the new symbolic logic and formal methods seized the spotlight (Rorty, 1979). Fascination with the new logic led Carnap and some other members of the Vienna Circle to reduce epistemology and any other philosophical topic worth saving to the logical study of language structure. Since methodology already was considered an application of epistemology, this development effectively reduced methodology to logic (including formal, inductive logic). According to Carnap:

Philosophy deals with science only from the *logical* viewpoint. *Philosophy is the logic of science*, i.e., the logical analysis of the concepts, propositions, proofs, theories of science, as well as . . . possible methods of constructing concepts, proofs, hypotheses, theories . . . . [Philosophy is the] logical syntax of the language of science. (1934, pp. 6, 9; Carnap's emphasis)

The next two sections critically review — all too briefly — the work of Popper and Lakatos on my three topics of methodology, heuristics, and rationality.

III. POPPER

A. *Methodology*. There is some truth in the claim that Popper is the greatest living, 19th-century methodologist; for Popper, as much as anyone after 1925, saved methodology in the old sense from near-extinction at the hands of formalists. Directly contrary to Carnap, Popper made the problem of the growth of knowledge the central problem of methodology, leaving epistemological foundationism no positive place at all.[6]

Much in Popper reminds us of Peirce: the fallibilism, the supremacy of method in defining scientific activity, the emphasis on problems and problem solving, the intellectual Darwinism, etc. But Popper retains a correspondence theory of truth (a true proposition is in correspondence to some other thing), whereas Peirce went further in (problematically) defining truth in terms of the inquiry process itself. We might say that Peirce's view of truth is entirely "formal" or "procedural" (Truth is whatever our method gives us, by convergence in the limit), whereas Popper retains a "substantive" concept of truth as independent of method.

Popper has written so much over the years that his position on methodology is not easy to pin down. On the one hand, he sometimes denies that there is a scientific method or methods. Popper is against method, although not in Feyerabend's radical way; for in such passages, Popper usually takes method to be a mechanical, algorithmic procedure or recipe for making new discoveries. On the other hand, in *Logik der Forschung* (Chap. II) Popper explicitly commits himself to the existence of methodology in the sense of a theory of the rules of the game of science. Nevertheless, in this weaker sense of 'method' there still is no method of discovery. Popper's rules fall on the side of justification (or, rather, "corroboration") rather than discovery.[7] Thus his methodology is schizophrenic, during the creative phase of research, anything goes, and an irrational element of inspiration is necessary; then, suddenly, at the second stage, logical rigor sets in and imaginative generation of new ideas ceases. The trouble is, rigorous testing is pointless unless there is some epistemic reason (by virtue of the way in which candidate hypotheses are produced) for thinking that the small subset of the myriad possible hypotheses to which testing is applied contains the truth, or something close to the truth. Popper explicitly denies that there is such a reason.

Actually, it is more accurate to say that Popper has *no* methodology of the research process, rather than half a methodology. To be sure, Popper often appears to discuss the research process rather than the logic of the finished product, the final research report. But there is no logic or algorithmic method of justification anymore than there is one for discovery. Obviously, the testing process is itself full of search and generation tasks — "discovery" tasks. There is no step-by-step recipe for criticizing and testing theories anymore than for creating them. For instance, there is no mechanical method for finding inconsistencies in a theory. Nor is there a recipe for generating novel predictions from a theory — and novel predictions are the *only* empirical evidence which Popper and his circle admit as corroborating evidence. Popper's sharp and invidious distinction between discovery and justification stands up only if Popper is concerned chiefly with the logic of the finished research report. But in that case, Popper has almost nothing to tell us about methodology of science as it concerns the *process* of inquiry.

Popper's Humeanism, Kantianism, and deductivism are apparent in a famous passage:

The initial stage, the act of conceiving or inventing a theory, seems to me neither to call for logical analysis nor to be susceptible of it. The question how it happens that a new idea occurs to a man — whether it is a musical theme, a dramatic conflict, or a scientific theory — may be of great interest to empirical psychology; but it is irrelevant to the logical analysis of scientific knowledge. This latter is concerned not with *questions of fact* (Kant's *quid facti?*), but only with questions of *justification or validity* (Kant's *quid juris?*)....

... [T]he method of critically testing theories, and selecting them according to the

results of tests, always proceeds on the following lines. From a new idea, put up
tentatively, and not yet justified in any way ... conclusions are drawn by means of
logical deduction. These conclusions are then compared with one another and with
other relevant statements, so as to find what logical relations (such as equivalence,
derivability, compatibility, or incompatibility) exist between them. (1959, pp. 31—2)

As the final sentence reminds us, Popper restricts logic of science to the
simple relations of deductive logic, logical deducibility, inconsistency,
and independence. Since he officially excludes inductive logical relations and relations of any other sort, he is, in this respect, more
restrictive even than Carnap. And unlike Peirce, Popper is not much
concerned with efficiency and the economy of research (Rescher,
1978). Only logically necessary relations are directly relevant to
methodology. As Lakatos (1970, p. 103) unintentionally points out,
Popper's view leaves method with only the slimmest logical resources:
"If all scientific statements are fallible theories, one can criticize them
only for inconsistency." This does not mean that Popper simply reduces
method to logic. Recall his official view that method consists largely of
rules which have the status of conventions adopted by human decision.
Since logical relations hold only between statements and not between a
statement and the world, even the most basic empirical observation
does not logically support a data *statement*. Which claims are accepted
as data statements (as opposed to theoretical claims) and which are
discriminated as *true* data statements are matters of community decision; they are *conventions*. The truth is not "given" by nature speaking
directly to us. The point of speaking of conventions here, against the
naturalistic view of method, is that scientific methods, while constructed
with an eye to scientific experience and fertility, are supposed to be
content-neutral and not theory laden. They are methods which presumably apply to all sciences at all stages of development. (Yet at this level,
Popper neglects to discuss the evolved social conventions governing all
rational argument, to which Dewey, Rorty, and Rescher are alive.)

Popper's Humeanism grounds his view of method as strongly normative and as content-neutral. Popper credits Hume with clearly distin-
guishing questions of validity (answers to the question *quid juris*?) from
questions of fact (answers to the question *quid facti*?). Subsequently,
Kant identified philosophical questions with the former rather than the
latter. That is, Kant awarded to philosophy the normative, nonempirical
task of demarcating and legitimating genuine sciences from other
projects — rather than the task of finding out how science "works" in

practice. Following Hume and Kant here leads Popper to overemphasize demarcation issues and to draw the *is-ought* distinction so sharply as to prevent him from accounting for the normative force of methodological rules in terms of their previous success or failure in scientific practice.

Stated more sympathetically, what Popper proposes is a *critical* methodology rather than a *constructive* one. Popper tells us that, until he showed otherwise, people thought that the only form of empiricism was inductivism. Since Hume (allegedly) demonstrated that constructive empiricism is untenable, precisely because it attempts to be constructive — to provide rules for discovery, for constructing true, general scientific claims — the failure of inductivism appeared to mean the failure of empiricism. Besides, constructive methodologies are necessarily foundational and authoritarian, Popper (1963) says (mistakenly). Meanwhile, Popper himself succeeded in developing a nonconstructive, fallibilistic, nonauthoritarian, critical empiricism which is not at all inductive; on the contrary, it is entirely deductive. In short, Popper showed how to be a Humean while remaining an empiricist. Or so Popper claims. That is why his view of science is schizophrenic. All rational considerations imply the existence of plausible, deductive arguments for the conclusion in question. Since there are no deductive, discovery arguments, Popper mistakenly thinks, the discovery process cannot be rational. But once a new idea is on the table for criticism, the whole nature of scientific activity abruptly changes, for predictions may be deduced. However, I insist on my earlier point. As a *process* of inquiry, criticism is also an active, constructive task, whence our talk of "searching criticism."

B. *Heuristics.* Given Popper's emphasis on problem solving as the essence of inquiry, he says amazingly little about how scientists actually do solve their problems, about heuristics. Popper frequently discusses the overall order of inquiry — problems first, next conjectures, and only then are factual data brought to bear, in a critical role; he argues (very questionably) that his methodology is more efficient and more systematic than inductivism (Popper, 1962, p. 51; 1963, p. 965); and he does have his "methodological correspondence principle" (Any adequate theory must yield its predecessor, to a first approximation). Yet, like Dewey, Popper is content with very general remarks and never gets down to detail. It is only his restriction of method and methodology to what all sciences have in common (deriving from his early concern to demarcate Science from everything else) that lend plausibility to his claim that scientific method is hypothetico-deductive to the core (Nickles, 1984b).

In point of fact, no completely general methodology can include strong heuristics, for the more powerful heuristics are knowledge-based.

... There is a kind of "law of nature" operating that relates problem solving generality (breadth of applicability) inversely to power (solution successes, efficiency, etc.) and power directly to specificity (task-specific information). (Feigenbaum, 1968.)

Popper's method of conjectures and refutations actually amounts to one of the weakest heuristics of all, generate and test, where "anything goes" at the generation stage. Contrary to Popper, then, there clearly *is* something between rigorous algorithms and psychological *Einfälle*, namely, heuristic problem-solving methods. Since heuristic methods (in a broad sense) are not only available but also of great importance in scientific research, in this respect Popper's non-view of method sets methodology back to a pre-Newtonian, pre-Cartesian, pre-Baconian level, in which new results are purely a matter of luck. The whole thrust of the 17th-century concern with method was to counter that view and to improve on the previous history of nondiscovery.

C. Rationality. Which methods are rational presumably depends on what science aims to achieve, and which aims are rational depends on which methods are available. Popper devotes much attention to the question of aims. His answer is that science aims at interesting, true theories — true theories with great depth of empirical content. Yet, ironically, his own "method of conjectures and refutations" or "method of trial and error," as he often describes it, obviously is inadequate to this task. The method does not tell us how to achieve this goal, how to recognize goal attainment once we have it, or even whether we are closer to the truth than before (Laudan, 1981a, p. 533; Stove, 1982). Moreover, according to Popper himself, it is most improbable that we shall ever attain the goal, in whole or in part. This being the case, why is truth important as a goal? Does he follow Descartes and Bacon in confusing truth with utility? Not likely, since Popper is contemptuous of practical considerations. Yet he writes:

Assume that we have deliberately made it our task to live in this unknown world of ours; to adjust ourselves to it as well as we can; to take advantage of the opportunities we can find in it; and to explain it, if possible ... with the help of laws and explanatory theories. *If we have made this our task, then there is no more rational procedure than the method of trial and error — of conjecture and refutation.* ... (1962, p. 51.)

But given that the number of possible theories on any topic is virtually boundless, it is hard to see how repeatedly refuting particular, bold proposals is likely to lead us to the truth or even to "adjusting ourselves" to the world as well as we can. (Popper admits that his method cannot explain the progress of science, though he would insist that no method can.) Popper's method seems a particularly inefficient, unfruitful way of proceeding in science (Grünbaum, 1978). The low cost of boldness (easy testability) is overbalanced by the negligible benefit — we learn almost nothing. Popper's method is so "hit or miss" that some writers, with considerable encouragement from Popper's own writings, have taken it for blind variation plus selective retention.[8] Ironically, Popper needs a rational, quasi-constructive account of "discovery" or problem solving in order for the critical side of his method to have much point.

D. *Critical appraisal of Popper.* Despite their attractive features, Popper's method and methodology have, in certain respects, contributed to the decline of methodological thinking in the 20th century. Popper declares that epistemology derives from scientific methodology rather than *vice versa*; yet Popper's methodology hinges on Hume's epistemology (as skeptically interpreted) — an epistemology that never concerned itself with genuine problems of real science (Laudan, 1981b) and that never figured seriously in methodology of science before Popper. Hume's form of the problem of induction figures centrally in Popper, while Popper's methodology, with its largely misplaced emphasis on boldness and novelty, gratuitously exacerbates the more serious, *scientific* problem of induction, the problem of the underdetermination of theory by observation. Popper fiddles while Rome burns. Popper says the immediate goal of inquiry is to solve important problems, but this (praiseworthy) stance scarcely justifies his emphasis on hypothetical boldness and on the production of esoteric, novel predictions (which almost becomes an end in itself) and his denying evidential status to anything else. Like the pragmatists, Popper defines science in terms of its method, yet he denies that there is a scientific method. Like the pragmatists, Popper adopts a problem-oriented account of method as problem solving, but again, he denies that there are any *methods* of problem solving.[9] Rather, he retains an irrationalist view of search and generation tasks, or at best employs only the very weakest heuristic, trial and error. Popper makes truth the goal of inquiry, but the methods his methodology espouses are demonstrably incapable of determining whether this goal is ever achieved or whether science is progressing toward it. The importance of achieving theoretical truth to the actual working of science is accordingly unclear. Hence, Popper's methodology is irrationalist. Given Popper's deductivism and commitment to a skeptical Humean position on empirical support, it is not surprising that his methodology consists mainly of a few, very general rules. It contains few resources to guide inquiry.

IV. LAKATOS

A. *Methodology.* For Lakatos (1968, 317ff) methodology, though not scientific practice itself, is, in a broad sense, "heuristics all the way down."[10] Lakatos terms scientific method "logic of discovery" or "heuristic," the application of which aims to increase knowledge. Methodology is the rational reconstruction of patterns of the growth of knowledge. Lakatos (1978, p. 101) explicitly distinguishes heuristic and epistemological considerations and criticizes the attempt to merge them as "justificationist." Lakatos's view is a radical reversal of the Carnapian identification of methodology with confirmation theory. The methodological problem of the growth of knowledge and the foundational problem of justification were the twin problems of classical epistemology. Since foundational epistemology turned out to be bankrupt, the methodological problem and *not* the problem of justification should be the focus of contemporary epistemology of science, says Lakatos. But Carnap and his followers doggedly pursued the latter path. They became so absorbed in hopeless, epistemological-foundational issues and in the technical problems surrounding them that they did "not even seem to have a word for what used to be called 'methodology', 'heuristic', or 'logic of discovery'" (1968, p. 326). Rather, they "expropriated" the term 'methodology' for the discipline of applied inductive logic. Lakatos agrees that problems of appraisal are important but thinks that all we need say about appraisal will fall out of a fallibilistic account of the growth of conjectural knowledge. In other words, for Lakatos, interestingly, appraisal itself is methodological — inseparable from, and, indeed, derivative from, methodology (pp. 328, 376).

Now it is true that Popper has championed the anti-Carnapian conception of methodology as the study of the growth of knowledge. Yet, given Popper's sins of commission and omission against discovery and heuristics, Lakatos faces the delicate task of grafting a genuine heuristic methodology onto Popperian hypotheticalism. In the last analysis, the graft does not seem to "take," for even a fallibilistic, heuristic methodology will be constructive, not merely critical-eliminative. While Lakatos both adds constructive elements to Popper and subtracts critical elements (e.g., greatly diminishing the importance of falsification), he never properly reconciles Popper's brand of hypotheticalism, in which *all* empirical support derives from successful novel predictions, with the central idea of constructive heuristics — that part of the warrant of a problem solution is the method by which it is generated.[11] The result is that Lakatos's methodology ends up being an unholy amalgam of informal heuristic ideas traceable back to Descartes and beyond, on the one hand, and of Popperian hypotheticalism on the other. Lakatos wrongly thinks that a Popperian sort of concern with the growth of knowledge is the only alternative to the untenable, classical discovery program.[12]

If Hacking's (1979) insightful review of Lakatos's work is correct, Lakatos's methodology is closer than Popper's to Peirce's pragmatic methodology. For while both Popper and Lakatos define science in terms of method rather than a body of substantive results, and while both in some sense base the warrant of scientific claims on the method used to produce them, Popper takes for granted the correspondence theory of truth, whereas Lakatos remains uncommitted to the view that a proposition is true by virtue of representing some external thing. As Hacking interprets him, Lakatos shares Peirce's post-Kantian aim of replacing *representation* by *methodology*. There is no external criterion of truth, whether humanly accessible or not.

Be that as it may, Lakatos rightly aims for a methodology that will fill the large gap between algorithmic rigor and Popper's trial and error. In the Introduction to *Proofs and Refutations*, Lakatos explicitly distinguishes his methodology of mathematics from dogmatic, ahistorical, formalist metamathematics and metalogic (he quotes Carnap as an example) and adopts an "informal," historical approach. Lakatos states that his approach will furnish an alternative to "the bleak alternative between the rationalism of a machine and the irrationalism of blind guessing" (p. 4) — between clocks and clouds, as it were. His methodology is historical in the further sense that the empirical support relation is not simply a logical relation between theory and data statements but depends also on whether or not the data was used to construct the theory (see Note 11).

B. *Heuristics*. In a broad sense, then, Lakatos conceives methodology as heuristics. To his credit, he gives far more importance to heuristics than any of the leading "general model builders" of science, Popper, Kuhn, Feyerabend, Toulmin, Laudan, etc. In my opinion, this change in the construal of methodology is of major significance.

My account implies a new criterion of demarcation between 'mature science', consisting of research programmes, and 'immature science', consisting of a mere patched up pattern of trial and error. (1970, p. 175, Lakatos's emphasis.)

In his early work, *Proofs and Refutations*, Lakatos apparently envisioned a general, content-neutral heuristics-cum-methodology of problem solving. Later, however, with Lakatos's development of the mature methodology of scientific research programs, the heuristics became explicitly content-specific: each research program possesses its own negative and positive heuristic, and the heuristics differ greatly from one program to another. Moreover, these heuristics are not just alternative sets of neutral rules or conventions for practicing the game of science.[13] Rather, the heuristic of a research program normally will express a basic metaphysical position. The heuristic amounts to a kind of *transducer* for converting metaphysical ideas into a progressive series of scientific theories.

Yet just when Lakatos has prepared himself to remedy the major epistemic defect in Popper's position, he apparently denies that heuristics carries epistemological weight! At least it remains unclear whether or not, and why, heuristics is *epistemically* important to inquiry. Lakatos (1970, p. 124) accepts Popper's view that classical, justificationist ideas of support are virtually inconsistent with predictionist and anti-ad-hocness remarks by Leibniz and others. Popper allegedly demolished all forms of justificationism, leaving the field to his predictive, anti-ad-hoc methodology. In their attempts to sharpen Popper's discussion of adhocness, however, Lakatos (1970, p. 175n) and Zahar (1973) are not content to say that a theory (or modification of a theory) is ad hoc if it contains no "excess" empirical content or (in another sense of adhocness) if none of its excess content is corroborated. They think it necessary to specify a third, heuristic sort of ad hocness. A theory is ad hoc in this third sense

if it is obtained from its predecessor through a modification of its auxiliary hypotheses which does not accord with the spirit of the heuristic of the programme. (Zahar, 1973, p. 101.)

Unfortunately, Lakatos and Zahar never clearly explain the epistemic rationale of this third requirement, that any non-ad hoc theory change must conform to the heuristic of the program. (This requirement would make any branching off into alternative research programs ad hoc, and Lakatos himself wants to encourage a certain amount of proliferation.) Given that scientific growth is a primary goal of Popperian-Lakatosian methodology, we can understand (even if we are not convinced of the need for) the first two ad hocness prohibitions. For the faster that a program churns out new results, the more likely that it will turn out well corroborated, interesting results. Thus a program with a stronger heuristic will tend to be epistemically superior (by Lakatosian standards) to any competing program with a weak heuristic. The heuristic therefore has an indirect, epistemic importance. But why the third prohibition — *unless the heuristic itself is thought to furnish pre-testing support*? For, *ex hypothesi*, in the third case we already have new results to check, even though not produced by the official heuristic. But why restrict science to new results produced by the official heuristic unless the heuristic imparts some epistemic weight to its products? Moreover, at some point, departure from a presumably exhausted heuristic may be the most fruitful thing to do — though not fruitful as defined by *this* programme-specific heuristic itself, of course. Lakatos is caught with general, content-neutral rules of the game on the one hand (his methodology of scientific research programs) and very specific heuristics on the other. There is nothing in between — making his position more Kuhnian than he intended.

C. *Rationality.* It follows from Lakatos's conception of methodology as the rational reconstruction of historical patterns of scientific growth that his methodology is backward-looking. Notoriously, Lakatos's methodology provides ways to appraise research programs but stops short of giving advice. This is astonishing, for the very idea of a method is the idea of something that guides inquiry, however fallibly; and the very idea of methodology is that of something that endorses specific methods as preferred directives for future behavior. The idea of a *heuristic* methodology which gives no advice is a contradiction in terms. Bluntly stated, Lakatos has no methodology. (Lakatos might cheerfully agree, but why then does he continue to describe his work as a 'methodology'?) So far as the future is concerned, Lakatos comes perilously close (as Feyerabend gleefully charges) to "anything goes." Lakatos's point seems to be an application of Hume: past patterns of growth and past successes of particular research programs furnish no reasons for thinking the future will be like the past. For, despite his emphasis on heuristics, Lakatos does not consider successful old heuristics to provide reliable advice about which means are suitable for reaching future goals. Heuristic promise and heuristic power never get translated into rules for action.[14] Hence, Lakatos provides no theory of rationality in the sense of something that furnishes reasons for and against future options. He inveighs against "instant rationality." Rationality judgments, like judgments about crucial experiments, are retrospective, historical judgments which afford no future guidance. His discussion of conventional decisions suggests that some sort of decision theory might replace traditional theory of rationality, at least where scientific methodology is concerned; but Lakatos does not develop this line of thought.

D. *Critical appraisal of Lakatos.* Despite his continuity with Popper, Lakatos is the first major philosopher of science in the contemporary period to make fairly detailed, content-dependent heuristics central to methodology. Yet he overlooks — or rather denies — the linkage of heuristics to rationality and therefore, like Popper, ends up disavowing methodology in any traditional sense. Lakatos's methodology of scientific research programs does emphasize problem solving, order of inquiry, and gain of efficiency over Popperian trial-and-error methods; and Lakatosian methodology is more historically conscious, detailed, and domain-specific than Popper's. Yet his haste to "rationally reconstruct" history in his own methodological image is the normative tail wagging the descriptive dog. Nor does Lakatos ever justify making novel prediction more important an epistemic criterion than problem-solving success (or, rather, making successful novel predictions of heuristically motivated changes the *only* criterion of problem-solving success), while heuristic derivation itself carries zero epistemic weight. Actually, Lakatosians reduce the importance of empirical testing (relative to Popper and other hypotheticalists) in favor of heuristic derivability of theories — but proceed to deny that heuristic derivability carries any epistemic weight whatever! This net loss in epistemic support is gratuitous and irrational. Finally, there is a strange disconnection of Lakatosian heuristics both from forward-looking guidance of inquiry and from epistemological warrant. This is a curious position for one who holds that the products of inquiry are, in part, warranted by the method of producing them.

V. METHODOLOGY: BETWEEN REALISM AND PRAGMATISM

What justifies adopting one method of investigation over another or, indeed, adopting any method at all? What is the source of the normative force of methodological pronouncements? So long as methodology is no more than a direct application of the rules of logic, this question reduces to the question of the justification of logic, which topic I cannot here pursue. If one has a more liberal conception of methodology as permitting conventional rules, the justification is likely to be *sociopolitical*, in part. As Popper conceives them, such rules are free decisions of the scientific community.[15] The rules presumably acquire their force in somewhat the way in which legislated laws do in a democracy — through general agreement of the community. (Of course, the perceived fruitfulness of the rules will be the crucial selling point, so this view of justification may be combined with others.) The community is free to adopt any logically consistent set of rules of the game of science which it wishes, subject to the constraint of the meta-rule that criticism and growth be maximized. This meta-rule (which is itself justified largely by its direct connection to the ends of inquiry and to epistemological fallibilism) virtually implies that the scope of conventional decision be minimized. This constraint is apparently severe and leaves little room for convention-making, since each Popperian argues that his or her variant of Popperian methodology is best on the ground that it opens up avenues of criticism not present in previous versions.

Another sort of justification of methods is *heuristic*. The justification here is empirical-scientific, not socio-political: because of the way the human brain and human society work, it is better (more fruitful, more efficient) to organize scientific investigations in certain ways and not others. Notice that the supporting empirical information concerns not the domain of scientific investigation (particle physics, population biology, etc.) but the cognitive psychology (memory capacity, learning capabilities, etc.) and sociology (e.g., the informal and institutionalized patterns of justification) of the community of investigators. Heuristics is concerned with economy of research. As in the literal economic sphere, economy of research becomes a necessary subject because of the scarcity of intellectual resources.

A third sort of justification of method is *pragmatic*. A method is justified to the extent that it works (or can be shown to work as well as a competing method). This idea can, of course, be combined with the last; but the methodological pragmatist stresses that the ultimate test of methods and methodologies is not whether they are in accord with abstract logical and epistemological principles or even with current, cognitive psychological and sociological theories but whether they work in practice. The justification is purely instrumental (teleological, purpose-directed), not theoretical. *Methodological* pragmatism is supposedly an improvement on *thesis* pragmatism, which holds that individual claims are epistemically justified by their utility. The trouble is, again, that the utility of a thesis is not coextensive with truth.[16]

Finally, there is the *realistic* type of justification. The pure realist argues that, since substantive, content-specific methodological rules are theory-laden, the rules are justified *only* to the extent to which the theories in question are (known to be) correct. (The parentheses indicate an ambiguity in the position.) Thus methods are justified to the extent that their substantive presuppositions are (known to be) true descriptions of reality and by that alone.[17] In contrast to heuristic justification, this sort of empirical justification of method pertains to the domain of investigation and not to the psychology of the investigator.

The pure realist and the pure methodological pragmatist positions are diametrically opposed on the direction of justification, the realist saying that methods are justified only if the theories they presuppose are true, the pragmatist retorting that the methodological process is what justifies the theoretical product. "[T]he rational legitimation of a method is not at all a question of *theoretical* considerations ... but is essentially *practical* ..." (Rescher, 1977, p. 4). To realism, the pragmatist will object that we *never* really know whether a theory is true, only how well it works in practice (practice including its predictive track-record): hence, all justification ultimately is pragmatic. Thus the realist, too, confuses truth and utility. Even if truth did guarantee utility, which it does not, we have no independent way of determining the truth. It is production by the right method (or satisfaction of methodological constraints) which justifies theories, not vice versa. Method itself is justified by the fact that *it* works in practice. And the better of two methods is the one which works better. What 'working better' means will vary from case to case. Sometimes we need high reliability. At other times, we want interesting problem solutions as rapidly as possible. Here efficiency of the methodology will be especially important — efficiency in searching the problem space or in generating scientifically interesting results.

Let us bring out more fully the motivation of the realist and pragmatist positions. The positivists and (for different reasons) the early Popper, wished to distinguish sharply between science and metaphysics. Yet even the positivists recognized that some metaphysical ideas guide scientific work. Their partial solution to this embarrassment was to reformulate untestable metaphysical principles as methodological claims. Instead of saying that every event in the universe has a cause, we make it a rule that no event is satisfactorily explained until a cause is found for it. Instead of saying that nature is simple, we say that inquiry begins with the simplest ideas and models and introduces complexity only as necessary. Substantive claims give way to regulative principles.

This solution to the embarrassment of metaphysics fails, the realist will reply. For the methodological rule will work well, and its success can be explained, if and only if the corresponding substantive presupposition is true. There is a commitment to metaphysical assumptions or theoretical truths after all. As the advent of quantum mechanics showed, the causality rule breaks down at a certain level: and it breaks down precisely where the classical ideas which formed its presupposition break down. Hence, the successes and failures of a methodological rules can be explained, and application of the rule justified, only in terms of the truth or falsity (and degree of accuracy, to the relevant approximation) of substantive claims about reality. Pope's poetic lines at the head of this paper are essentially correct. The more useful, the more detailed methodological rules and procedures are not theory neutral. They are domain-, theory-, and problem-context specific. Even if one adopts instrumental goals for science rather than truth, this point still applies. For, again, a method "works" only if it is consonant with reality.

The pragmatist rejects the realist position as too extreme. A good method is one which works in practice and one which fits human capabilities for inquiry, regardless of the way the world is otherwise. This (and not the truth about the physical universe) is why, in physical science and elsewhere, we require objectivity in the sense of inter-subjectivity. This (and not the belief that the world is simple) is why good methodologies adopt appropriate versions of the rule, Simplicity First! By starting with simple models and analyzing the nature and sources of *their* failure, we learn much more than if we begin with very complex models and attempt to analyze their failures. Hence, it is a mistake to say that the success or failure of method is explainable only in terms of the truth of its substantive presuppositions. In fact, we often learn more rapidly by starting from *contrary-to-fact* assumptions about the world, rather than the most realistic assumptions. This just illustrates how a premature and priggish concern for truth can block the road to inquiry. Methods also have to fit the information available to us, regardless of whether that information discloses universal truth. That is part of the point of Simon's bounded rationality, and that is the point of the old joke about the man searching for his house key around the lamppost one night. Asked a passerby, "How do you know that you lost your key here in the light?" "I don't," he replied. "But if it isn't here, I have no chance of finding it!"

This story also hints at the large topic of economy of research — weighing the costs against the benefits of acquiring new information, pursuing one or multiple lines of research instead of others, etc. The economy of research is an essential part of methodology, yet cost-benefit considerations are not themselves epistemically relevant to the acceptance of knowledge claims and methods (Rescher, 1978). More-over, if the adequacy of method depended entirely on the truth about the domain being studied, then there would be no room for disagreement concerning research strategy (risky vs. conservative strategies, etc.) among those in agreement about the nature of the domain.

Pure realism is clearly untenable, but so is pure methodological pragmatism. Each wrongly neglects what the other emphasizes, and both neglect the heuristic justification of method based on cognitive studies of human and machine learning. Even in the case of game-like search problems which are better defined than most serious scientific problems, which search strategy we adopt (depth first, breadth first, hill climbing, best first, etc.; see Winston 1977, Chap. 4) will depend upon our warranted assumptions about the structure of the search space (i.e., about reality) as well as the extent of our information, resources, our psychology, sociology, and love of risk. Clearly, it is a mistake to make either theories or methods entirely parasitic on the other. There is an "interaction" or accommodation of one to the other which is part of "learning to learn." "Each chief step in science has been a lesson in logic," but the improved logic in turn has provided a stronger warrant for scientific claims. Much of the domain-specific methods available today were inconceivable prior to the theoretical maturity of those domains. And those methods in turn afford a far more efficient production of better warranted results than either classical inductive methods or the "trial and error," hypothetico-deductive methods which officially replaced them (Nickles, 1984b).

I conclude that the only reasonable position is to recognize the mutual interdependence of theory and method. We must reject both extremum positions — "methodological realism" and "methodological pragmatism" (as I have characterized them) — and instead steer between them. Obstacles remain for this "in between" position (Rescher, 1977, Chaps. 4, 7). For instance, is it not circular to consider successful theories and successful methods, each helping to warrant the other? And how about Hume's problem: why should methods successful in the past continue to succeed in future applications? A brief retort to the Humean objection (a pragmatically grounded, negative induction!) is that every methodology which has taken Hume's problem of induction too seriously has impoverished itself to the point of infertility. The fact that Popper and (to a lesser extent) Lakatos make Humean empiricism the basis of their methodologies of science means that, contrary to their advertisements, they, after all, make methodology derivative from epistemology. Indeed, Hume's form of the problem can be ignored without loss by practical methodology (the how-to-do-it component of methodology). For example, Hume's problem of induction is not a serious topic in books on experimental design. As for the circularity, the circle surely is virtuous and not vicious. As we learn to learn we must pick ourselves up by our bootstraps, but not all bootstrapping procedures are impossible (Nickles, forthcoming b, §§6 and 7).

VI. CONCLUSION

I shall conclude by outlining my own position on methodology in the form of twenty-three theses. I have not argued for all of these points in this paper; indeed, I have not even discussed some of them. But they do seem to me to form a coherent and attractive position.

(1) Methods are more or less useful means for reaching ends. (2) Methodology is the comparative evaluation of competing methods of investigation. Metamethodology is the comparative evaluation of methodological criteria. (3) Accordingly, the problems of methodology are descriptive, critical, and advisory (Reichenbach, 1938, Chap. 1). (4) In particular, methodology attempts to determine the most efficient (maximizing) or sufficiently efficient (satisficing) means of reaching goals. (Paradoxically, it may not be efficient to search for the most efficient means to an end.) (5) Since heuristics is the field directly concerned with this task, heuristics is central to methodology and is not a dispensable, psychological crutch vis à vis logic. (6) In fact, heuristic considerations can carry epistemic weight. (7) Not all goals are epistemic, even in science. (8) And methodology is concerned with the process of inquiry, not only the product. (9) Thus methodology is theory of inquiry rather than logic or epistemology in any narrow sense. (10) Failure to find workable methods for reaching goals provides a way to criticize methodologically the choice of goals and to work an accommodation between available methods and realizable goals: methodology goes beyond the instrumental justification of determining the best means to reach a predetermined goal. (11) Thus competing methods and methodologies can be rationally appraised from external points of view. (12) The normative force of methodology depends in part on the fact that it endorses those methods which actually have worked as effective problem solvers in the past. Certain methods are preferred to others because they work better. (13) Which methods do work better must be determined by an extensive examination of scientific practice, from study of human cognitive psychology and sociology, and from the economy of research, and not from general, abstract, epistemological principles alone. (14) Hence, attention to the history of successes and failures of various methods (part of the descriptive task) is of utmost importance to methodology. (15) But methodology is prospective as well as retrospective. While rooted in and legitimized by the past, it looks to the future. Any methodology worthy of the name provides guidance to inquiry — advice as well as appraisal. (16) This guidance need not take the form of algorithms, nor need it be failsafe: usually it will be heuristic. (17) Nor is methodology exhausted by a set of rules. (18) Thus, again, the resources of methodology extend far

beyond deductive logical relationships and general, conventional rules. (19) And hence methodology often presents alternatives requiring decisions, which may depend on the risks that one is willing or able to take. (20) As (13) suggests, a useful methodology will go beyond the vapid generalizations characterizing all scientific activity (and, indeed, almost all intellectual activity) which have dominated recent philosophy of science. This means that methodology no longer will be a single, unitary subject but will, at the more interesting levels of detail, break down into domain-and context-specific rules, practices, and advice. (21) Against pure methodological pragmatism, a detailed methodology is heavily theory-laden and research-program-laden. The success of the methods it endorses is dependent on the success of the theories it presupposes. (22) But against pure realism, there is a strong reverse dependence of successful theory on successful method. Much else besides true theoretical presuppositions contributes to the success of a method. (23) In any case, methodology must be fallibilistic and naturalistic in the sense that it employs no "external" or supernatural criterion of truth (or whatever) not actually available to natural investigators such as human beings.

A good example of this conception of methodology in action is William Wimsatt's (1980) continuing work on the units of selection controversy in evolutionary biology. Wimsatt's focus is on reductionistic research strategies and their systematic, heuristic biases. His work well illustrates how the descriptive, critical, and normative tasks of method-ology intertwine. Some encouraging signs for the future are the increased interest of philosophers of science in the details of various historical research strategies, in artificial intelligence, and in cognitive psychology and sociology. From the pragmatic and heuristic perspec-tive, these wider studies do not betoken the transformation of norma-tive methodology into a merely descriptive enterprise. On the contrary, it is these investigations which furnish the empirical basis for normative claims. As the Peirce aphorism at the head of the paper implies, methodology is science teaching by example.

University of Nevada, Reno

128 THOMAS NICKLES

NOTES

[1] This paper contains only minor changes from the Italian version. For recent developments of my views, see Nickles (1985, 1986) and (forthcoming a, b, and c).

[2] Pope, *Epitaphs*, Intended for Sir Isaac Newton; Popper (1959, Chap. II): Peirce (1877, §1).

[3] Unfortunately, I did not read Rescher (1977, 1978) until this paper was virtually finished. Rescher develops several of "my" themes in much greater detail, although I naturally disagree at some points. For one thing, I conceive method in a more nominalistic, local, domain-specific way, as problem-solving techniques.

[4] Popper's fallibilistic rejection of perfectionism in epistemology and his dismissal of overprecision are steps in this direction. However, Popper retains an obsession with truth uncharacteristic of the pragmatists. Popper (1957) says that any methodology which fails to make truth the direct goal of inquiry blocks criticism and hence blocks the road to inquiry. An opponent may retort that since direct truth-seeking is a kind of optimizing strategy, it is Popper's methodology which blocks inquiry.

[5] However, formal logic and mathematics figured more prominently in Peirce, who made important contributions to these fields, than in Dewey, whose view was more thoroughly socio-historical.

[6] Lakatos claims that

> The main feature of classical empiricism was the domination of the logic of justification over the logic of discovery .... On the other hand, in Popper's treatment of growth without foundations the logic of discovery dominates the scene. (1968, p. 324n.)

This claim would be outrageous — the very reverse of the truth — were it not for Lakatos's sweeping equation, *logic of discovery* = *heuristics* = *philosophical reconstruction of the growth of knowledge* = *methodology*.

[7] Sometimes Popper includes even discovery within scientific methodology. Suppose, he says, that a clairvoyant produces a book, through automatic writing, that years later is found to be identical with a scientific work just published by a scientist who knew nothing of the clairvoyant's work. "We shall have to say that the clairvoyant's book was not when written a scientific work, since it was not the result of scientific method" (1962, pp. 218—19). It is hard to reconcile this curious passage with Popper's falsifiability criterion of demarcation and with his views on discovery.

[8] For example, see Popper (1965, p. 247), for whom discovery and methodology are either like clouds or else like clocks, with no intermediate possibility. Peter Skagestad (1978), Paul Thagard (1980), Marcello Pera (1981, 1982), and Robert McLaughlin (1982) all have criticized Popper's method for inefficiency and have attempted to remedy the problem by a tighter linkage of discovery (in the sense of original conception) and justification. For fuller discussion, see Nickles (1985) and (forthcoming a) and Rescher (1977, 1978). On the other hand, Campbell (1974 and elsewhere) develops the "blind variation plus selective retention" theme in a more sophisticated way than Popper. Like the just-mentioned authors, I am guilty of having misinterpreted Campbell in my previous publications.

[9] Perhaps Popper's claim is merely that there is no global, algorithmic method of discovery. Only in this traditional sense of the scientific method is there no logic of

discovery. Perhaps he allows that there are local discovery methods, but he denies that these are philosophically interesting? I do not think this view is ascribable to Popper but other critics of discovery programs hold it. In response, I ask why a view has to be globally true to be epistemologically interesting..

[10] This aphorism is William Wimsatt's, not Lakatos's. It applies to Lakatos only if we take 'heuristics' in his excessively broad sense.

[11] In fact, Lakatos and his students (Worrall, 1978; Zahar, 1983), never really justify their Whewell-Popper position of giving predicted data special epistemic weight and giving *no* other information any epistemic weight at all. They accept, almost without argument, the central dogma of hypothetico-deductive, consequentialist theories of empirical support: *empirical support = empirical evidence = successful test results = successful predictions = true empirical consequences of the theory tested* (plus auxiliary assumptions).

[12] See Nickles (1984a, 1985, and forthcoming a) for more on tensions within Lakatosian heuristics and for my attempt to sketch an alternative view.

[13] At this general level, Lakatos does seem to retain a single set of rules — his general methodology of scientific research programs — and does not allow proliferation, with some programs following his methodology, others Popper's, still others various versions of inductive justificationism, etc.

[14] Urbach (1978) provides a more forward-looking account of heuristics; but he, too, fails to explain why heuristics is epistemically important.

[15] Lakatos (1968) distinguishes dogmatic and naive from sophisticated and methodological conventionalism.

[16] See Rescher (1977) for an enlightening discussion of these issues. Rescher, too, recognizes but underplays the "reciprocity," the mutual justification, of methods and results. Some lingering questions are: Why is methodological pragmatism a more reliable indicator of truth than thesis pragmatism? Does not the content-ladenness of problem-solving methods spoil the neat separation of substantive thesis and method on which Rescher's justification strategy depends? At the frontier of a mature science, with content-specific methods, does not the content change so quickly and so profoundly that method is not sufficiently stable for the method to license its products? Is Rescher too much concerned with truth for a pragmatist?

[17] This entails Donald Campbell's (1974) view that at the frontiers of research, where we allegedly know nothing of what is beyond, heuristic method collapses into the weakest possible heuristic: generate and test, blind variation plus selective retention, conjectures and refutations. On my previous injustice to Campbell, see note 8.

BIBLIOGRAPHY

Campbell, Donald. (1974). "Evolutionary Epistemology." In P. Schilpp, ed. *The Philosophy of Karl Popper*. Library of Living Philosophers. LaSalle, IL: Open Court. Pp. 413—463.

Carnap, Rudolf. (1934). "On the Character of Philosophic Problems." *Philosophy of Science* 1, 5—19. As reprinted in *Philosophy of Science* **51**, (1984): 5—19.

Carnap, Rudolf, (1935). *Philosophy and Logical Syntax*. London: Kegan Paul.

Clarke, Desmond. (1982). *Descartes' Philosophy of Science*. University Park, PA: Pennsylvania State University Press.

130   THOMAS NICKLES

Dewey, John. (1938). *Logic: The Theory of Inquiry.* New York: Holt.
Feigenbaum, E. A. (1968). "Artifical Intelligence: Themes in the Second Decade." *Proceedings IFIP68 International Congress, Final Supplement.* Edinburgh.
Feyerabend, Paul. (1975). *Against Method.* London: New Left Books.
Garber, Daniel. (1983). "Semel in Vita: The Methodological Background to Descartes's Meditations," read at University of Pittsburgh.
Gaukroger, Stephen, ed. (1980). *Descartes: Philosophy, Mathematics and Physics.* Sussex: Harvester Press.
Gilbert, Neal. (1960). *Renaissance Concepts of Method.* New York: Columbia University Press.
Grünbaum, Adolf. (1978). "Popper vs. Inductivism." In Radnitzky and Andersson, (1978). Pp. 117—142.
Hooker, Clifford. (1977). "Methodology and Systematic Philosophy." In R. Butts and J. Hintikka, eds. *Basic Problems in Methodology and Linguistics.* Dordrecht: D. Reidel. Pp. 3—23.
Kuhn, Thomas. (1962). *The Structure of Scientific Revolutions.* Chicago: University of Chicago Press. 2nd edition, enlarged, 1970.
Lakatos, Imre. (1968). "Changes in the Problem of Inductive Logic." In *The Problem of Inductive Logic.* Amsterdam: North-Holland. Pp. 315—417.
Lakatos, Imre. (1970). "Falsification and the Methodology of Scientific Research Programmes." In I. Lakatos and A. Musgrave, eds., *Criticism and the Growth of Knowledge.* Cambridge: Cambridge University Press. Pp. 91—195.
Lakatos, Imre. (1976), *Proofs and Refutations.* Cambridge: Cambridge University Press. Originally published as a series of articles in *The British Journal for the Philosophy of Science* 14, (1963—64).
Lakatos, Imre. (1978). "The Method of Analysis-Synthesis." In *Philosophical Papers,* Vol. 2. Cambridge: Cambridge University Press. Pp. 70—103.
Laudan, Larry. (1977). "The Sources of Modern Methodology." In R. Butts and J. Hintikka, eds. *Historical and Philosophical Dimensions of Logic, Methodology and Philosophy of Science.* Dordrecht: D. Reidel. Pp. 3—19. Reprinted with changes in Laudan (1981b), pp. 6—19.
Laudan, Larry. (1978). "Ex-Huming Hacking." *Erkenntnis* 13: 417—35. Reprinted with changes as "Hume (and Hacking) on Induction," in Laudan (1981b), pp. 72—85.
Laudan, Larry. (1981a). "The Philosophy of Progress . . . ," in *PSA 1978,* Vol. 2, pp. 530—547.
Laudan, Larry. (1981b). *Science and Hypothesis.* Dordrecht: D. Reidel.
March, J. G. and H. A. Simon. (1958). *Organizations.* New York: John Wiley.
McLaughlin, Robert. (1982). "Invention and Appraisal." In McLaughlin, ed. *What? Where? When? Why?* Dordrecht: D. Reidel. Pp. 69—100.
Nickles, Thomas. (1984a). "Positive Science and Discoverability." *PSA 1984,* Vol. 1, pp. 13—27.
Nickles, Thomas. (1984b). "Scoperta e Mutamento Scientifico" ("Discovery and Scientific Change"). *Materiali Filosofici* 8: 7—27.
Nickles, Thomas. (1985). "Beyond Divorce: Current Status of the Discovery Debate." *Philosophy of Science* 52: 177—206.
Nickles, Thomas. (1986). "Remarks on the Use of History as Evidence." *Synthese* 69: 253—266.

Nickles, Thomas. (forthcoming a). "Lakatosian Heuristics and Epistemic Support." *British Journal for the Philosophy of Science*, in press.

Nickles, Thomas. (forthcoming b). "Twixt Method and Madness." In N. Nersessian, ed. *The Process of Science*. The Hague: Martinus Nijhoff, in press.

Nickles, Thomas. (forthcoming c). "Questioning and Problems in Philosophy of Science: Problem-Solving Versus Directly Truth-Seeking Epistemologies." In M. Meyer, ed. *Questions and Questioning*. Berlin: W. De Gruyter, in press.

Nickles, Thomas. (forthcoming d). "Professional Philosophy of Science: Some 'Sociological' Reflections."

Peirce, Charles. (1877). "The Fixation of Belief." *Popular Science Monthly* 12: 1—15; reprinted in C. Hartshorne and P. Weiss, eds. *Collected Papers of Charles Sanders Peirce*, Vol. 5. Cambridge: Harvard University Press, 1934, pp. 385—387.

Pera, Marcello. (1981). "Inductive Method and Scientific Discovery." In M. D. Grmek, R. S. Cohen, and G. Cimino, eds. *On Scientific Discovery*. Dordrecht: D. Reidel. Pp. 141—165.

Pera, Marcello. (1982). *Apologia del Metodo*. Rome: Laterza.

Polanyi, Michael. (1958). *Personal Knowledge*. Chicago: University of Chicago Press.

Popper, Karl. (1934). *Logik der Forschung*. Vienna: Julius Springer.

Popper, Karl. (1945). *The Open Society and Its Enemies*. London: Routledge & Kegan Paul. As reprinted by Harper & Row, New York, 1962.

Popper, Karl. (1957). "Philosophy of Science: A Personal Report." In C. A. Mace, ed. *British Philosophy in Mid-Century*. London: Allen & Unwin. Pp. 155—191. As reprinted with the title, "Science: Conjectures and Refutations." In Popper's *Conjectures and Refutations*. New York: Basic Books, 1962, pp. 33—65.

Popper, Karl. (1959). *The Logic of Scientific Discovery*. New York: Basic Books; English translation of Popper (1934).

Popper, Karl. (1963). "Science: Problems, Aims, Responsibilities." *Federation Proceedings, Federation of American Societies for Experimental Biology* 22, Part 1, pp. 961—972.

Popper, Karl. (1965). "Of Clouds and Clocks." As reprinted in *Objective Knowledge*. Oxford: Oxford University Press, 1972, pp. 206—255.

Radniztky, G., and G. Andersson, eds. (1978). *Progress and Rationality in Science*. Dordrecht: D. Reidel.

Reichenbach, Hans. (1938). *Experience and Prediction*. Chicago: University of Chicago Press.

Reitman, Walter. (1964). "Heuristics, Decision Procedures, Open Constraints, and the Structure of Ill-defined Problems." In M. W. Shelly and G. L. Bryan, eds. *Human Judgments and Optimality*. New York: John Wiley. Pp. 282—315.

Rescher, Nicholas. (1977). *Methodological Pragmatism*. Oxford: Blackwell.

Rescher, Nicholas. (1978). *Peirce's Philosophy of Science*. Notre Dame, Indiana: Notre Dame University Press.

Rorty, Richard. (1979). *Philosophy and the Mirror of Nature*. Princeton: Princeton University Press.

Sakellariades, Spyros. (1980). *The Role of Evidence in Descartes' Scientific Method*. Ph.D. dissertation, University of Pittsburgh Press.

Sakellariades, Spyros. (1982). "Descartes' Use of Empirical Data to Test Hypotheses." *Isis* 73: 68—76.

Simon, H. A. (1957). *Models of Man.* New York: John Wiley.

Skagestad, Peter. (1978). "Taking Evolution Seriously: Critical Comments on D. T. Campbell's Evolutionary Epistemology." *The Monist* **61**: 611—621.

Stove, David. (1982). *Popper and After.* Oxford: Pergamon press.

Thagard, Paul. (1980). "Against Evolutionary Epistemology." *PSA 1980*, Vol. 1, pp. 187—196.

Toulmin, Stephen. (1972). *Human Understanding.* Princeton: Princeton University Press.

Urbach, Peter. (1978). "The Objective Promise of a Research Programme." In Radnitzky and Andersson (1978), pp. 99—113.

Wimsatt, William. (1980). "Reductionistic Research Strategies and their Biases in the Units of Selection Controversy." In T. Nickles, ed. *Scientific Discovery: Case Studies.* Dordrecht: D. Reidel. Pp. 213—259.

Winston, Patrick. (1977). *Artificial Intelligence.* Reading, MA: Addison-Wesley.

Wittgenstein, Ludwig. (1964). *Remarks on the Foundations of Mathematics.* Oxford: Blackwell.

Worrall, John. (1978). "The Ways in Which the Methodology of Scientific Research Programmes Improves on Popper's Methodology." In Radnitzky and Andersson (1978), pp. 45—70.

Zahar, Elie. (1973). "Why Did Einstein's Programme Supersede Lorentz's?" *British Journal for the Philosophy of Science* **24**: 95—123; 223—62.

Zahar, Elie. (1983). "Logic of Discovery or Psychology of Invention?" *British Journal for the Philosophy of Science* **34**: 243—261.

PART II

RATIONAL SCIENTIFIC CHANGES

JOSEPH C. PITT

GALILEO AND RATIONALITY:
THE CASE OF THE TIDES

Galileo's theory of the tides occurs in Day Four of his *Dialogue on the Two Chief World Systems*.[1] The general interpretation of its placement at the end of the *Dialogue* is that it is Galileo's "crucial experiment" in defense of the Copernican system [Drake 1978; Koyré 1939; Shea 1972]. By assuming the motions of the earth, an explanation would finally be forthcoming for the motion of the tides. Unfortunately, it is a strange theory, adducing a primary and some secondary causes for the tides; the primary cause is insufficient to explain the data, necessitating the secondary causes. But it is the primary cause which invokes the earth's motions; hence, it appears that this is a fairly weak defense of Copernicus. This suggests Galileo's strategy in using the tides was flawed.[2]

Despite the apparent failure of Galileo's gambit in using the tides to defend Copernicus, I have argued elsewhere that the choice of the problem of the tides was rational [Pitt 1982]. In the tides, Galileo identified a research problem, the solution to which he saw as essential to constructing a new understanding of the structure of the universe consistent with the Copernican model.

Given that the choice of the problem of the tides has been shown to be a good one, in this paper I argue, in addition, that Galileo's solution and the role he assigned to the argument presenting the solution were also rational. The crux of the case rests on showing that for Galileo it was rational to accept a conclusion if that conclusion could be used in an explanation of some physical phenomenon. In this sense then, Galileo was an explanationist. For an explanationist, the justification for accepting an hypothesis is the demonstration of its explanatory power.[3] Since the key to understanding rationality in general is to understand what is accepted as justification, the clue to understanding the rationality of Galileo's solution to the problem of the tides lies in understanding what he took to be an explanation. Furthermore, Galileo's committment to explanation as the basis for justification is so strong that even if we were to accept the standard view of the structure of the *Dialogue*, we would be able to conclude that Galileo thought he had

135

J. C. Pitt and M. Pera (eds.), Rational Changes in Science, 135–153.
© 1987 *by D. Reidel Publishing Company.*

provided a justification for the Copernican view because he believed he had produced an explanation for the tides. Since the tides were a phenomenon which had proved troublesome for previous accounts, if Galileo could show, even reasoning *ex suppositione*[4] (which if my argument here is correct, turns out to be a very weak form of explanation for him), that on the hypothesis of the earth's motion a defensible explanation of the motion of tides could be constructed, that would have been a major success in his mind. But, while I argue that Galileo was an explanationist in matters of justification, and that he believed the theory of tides was justified, I also want to suggest that more is going on here than merely a defense of Copernicanism.

The key point is Galileo did not just have an explanation for the tides. The issue turns on what he counted as an explanation. In his conception of what a proper explanation of physical phenomena entailed we can see the beginnings of a philosophy of science, aspects of which are in force even today. On the view presented here, the real importance of the *Dialogue* is to the extent that Galileo could be described as having one, the *Dialogue* contains the heart of his philosophy of science. In particular, Galileo devotes the first two days of the *Dialogue* to establishing the adequacy of his conception of explanation as justificatory, as well as developing a new criterion of evidential support. The *Dialogue*, therefore, is a more complicated and sophisticated work than many think.

The current popular view, much of the credit for which is due to Feyerabend (1975), is that it was intended by Galileo as a work of propaganda for the educated layman and that it contains little of genuine scientific significance. But there *is* scientifically significant material here. And while it is true that few empirically significant discoveries are revealed in the *Dialogue* beyond Galileo's Law of falling bodies, what counts as scientifically significant extends beyond mere facts. Changes in methodology, criteria of evidence and explanatory relevance are also significant for science. Some of these are philosophically significant components of a given science; but since they also make a difference in the practice of scientists, they are scientifically significant. Viewing the *Dialogue* in this light, as a treatise in the philosophy of science, we can focus on Galileo's conceptualization of the problems he sought to solve. If we read the *Dialogue* as an articulation of Galileo's view of justification as explanation, and simultaneously as a defense of his approach to explanation, we can account for what, on more standard interpretations, appear to be tensions and conflicts within Galileo's work.

Needless to say, this is not a standard account of Galileo's philosophy of science. It conflicts in many ways with more orthodox views such as Ernan McMullin's (1978). In part, then, the success of my view will be a function of having shown first that traditional views, here typified by McMullin's exposition, fail to do justice to Galileo. This is the task of part one. In part two, I restrict the general discussion to a more specific account of Galileo's justification of his theory of the tides in the *Dialogue*. Part of this section is devoted to an analysis of the structure of the *Dialogue*. I conclude with a discussion of some problems associated with explanationism and Galileo's solution to those problems.

I. SCIENCE AND EXPLANATION

There are many conflicting themes and pressures in Galileo's work. These lead to a variety of questions, some more important than others. Here is a sample of the types of queries that have been raised. Is Galileo a neoplatonist, the last of the Aristotelians, the father of modern science, an experimentalist, an inductivist, a proto-positivist, a physicist, an engineer, a good Catholic, the first Defender of Scientific Truth against Church Dogma, an egomaniac, a man obsessed with caring for his family? The list, like the road, goes on and on. Furthermore, one cannot but be struck by the apparent incompatibility of many of these questions when taken together. And yet, McMullin, in his paper entitled "The Conception of Science in Galileo's Work", attempts to weave almost all of the above into a coherent view. As a unifying principle McMullin urges the thesis that Galileo was warring with himself over the most adequate view of science. As McMullin sees it, the competitors are (1) science in the light of the demonstrative ideal of the Greek tradition, and (2) what McMullin characterizes as the "retroductive notion of science," which is reasoning backwards from what one wants to explain to the most likely hypothesis "eliminating all hypotheses save one." [McMullin 1978, p. 227.]

Nevertheless, despite the effort to accomodate these different strands of thought, there are problems with McMullin's account. To begin with, there is a slide throughout his presentation from the expressed topic, i.e. Galileo's conception of science, to claims about what Galileo was

offering as an explanation. I take these to be different and explain why
below. There is also a problem with McMullin's use of the texts. I
cannot deal with this in the detail it needs, but here is one worry. In his
discussion. McMullin bounces back and forth over the *Dialogue*,
Discourse on Floating Bodies, *Letters on the Sunspots*, *Letter to the
Grand Duchess Christian*, *The Assayer*, and *The Starry Messenger* and
Two New Sciences without regard to their order of composition and the
length of time they cover in Galileo's life — over 34 years.[5]

The implication of McMullin's approach is that Galileo's conception
of science is to be seen as remaining constant throughout his working
career, or to use McMullin's language, that for Galileo the conflict
between the geometric and retroductive ideals of science is one of long
standing. I am not convinced. More precisely, I cannot agree with
McMullin's claim that this alleged tension reveals anything about
Galileo's conception of science. Not only was there no such tension, but
I doubt that Galileo had a conception of science relevant to any
concern of ours. This can be seen once we clarify the relation between
science and explanation for Galileo.

The issue of Galileo's conception of science cannot be settled by
appeals to what we or McMullin perceive to be Galileo's efforts at
explanation. For this in part begs the question. It assumes that we can
deduce Galileo's account of science from his views on explanation. This
is strange for two reasons. First, since on McMullin's theory Galileo
had two conceptions of science going, it is not clear how to derive these
from Galileo's singular theory of explanation. Second, even if we could
derive these two notions of science from Galileo's theory of explanation, it assumes, in that very derivation that the issue of conceptualizing
"science" was in some sense, if not settled, at least narrowed to these
alternatives for Galileo. But, in any relevant sense there was no
"science" yet and, furthermore, Galileo really had no idea of what
science would develop into. Hence, it seems premature to speak of
Galileo's conception of science. And this is not a problem peculiar to
Galileo. It was endemic to the times. For example, in speaking of the
development of the Royal Society in England and its rejection of the
past as its members tried to construct a new theory of experimental
science, R. S. Woolhouse notes that,

... it must be realized, moreover, that the apparently simple fact that the natural
science of later centuries is a product of that time has a hidden complexity to it. It is not
as though there always was a clear conception of what such a science would be like and

as though all that was lacking was a success in producing it. It is rather that the very idea of a body of knowledge about the world of the sort we have now, the very idea of a natural science, was being forged at the time. [Woolhouse 1983, p. 37.]

Galileo did have some views, let us call them intuitions, about where the important problems lay and what had to be done to solve them. Shea and McMullin point to a fine example. Galileo may have sensed that a general and unified theory of motion adequate to solving problems of both planetary and terrestrial motion was needed if any defense of Copernicus was to succeed. But, on the surface, it appears that he did not have a clue as to how to produce that theory. For one thing, such a theory would require some prior conception of the relevant kinds of evidential support appropriate to such a general theoretical framework. This, in turn, would presuppose at least a sketch of the sorts of experimental activities that would qualify as providing such evidence. At best Galileo could only hint at what was necessary. Thus, Galileo points to a direction for future research when, for example, he attempts to argue for "the similarities between the moon and the earth" in *The Starry Messenger* and the *Dialogue*, thereby articulating a sense of what was needed to discuss the heavens, namely a theory which abrogated the distinction between the celestial and terrestrial domains and allowed such comparisons in the first place. But to point to what is needed is not the same as providing criteria of adequacy for a successful theory. Likewise, to observe that given a unified theory of motion for celestial and terrestrial events, we could explain a large number of phenomena currently unexplainable, is not to say what constitutes science.

So let us see if instead of creating something *we* want to call science for Galileo, we can be a bit more successful in characterizing what *Galileo* believed he was doing and why. Beginning in 1613, Galileo felt secure enough about Copernicanism to publicly express hope for its success. Mundane as this observation is, its consequences have rarely been fully explored. For one thing, accepting Copernicus' view presented Galileo with a number of problems.

For example, what is going to count as evidence? How are we to explain the causes of motion when the grounds for the only fully developed theory of motion, Aristotle's, are no longer accepted by Galileo? In rejecting Aristotle's theory of motion Galileo was also rejecting an entire mode of explanation. But this is not to say that he rejected everything in the broader Greek tradition. For example, whatever else Galileo did, in the tradition of Archimedies he relied on demonstrative proof, paradigmatically Euclidean geometric proof, for conclusive\ presentation of his points. Furthermore, as we shall see, he urges the use of such methods in place of arguments from discursive definitions in the manner of Aristotle's latter-day followers.

But, given that Galileo is (a) Archidemean inclined and (b) rejects Aristotelean methodology, two problems arise. First, it is not all clear what he believed he had accomplished once he produced a geometric proof. Second, if that mode of demonstration was the only important one, why did he continually accompany his geometric proofs with examples drawn from common terrestrial experience? Let us press this point by putting the question boldly. In what sense is a proof an explanation? To see the question more clearly note that proof is primarily a formal or syntactic notion for Galileo, while explanation requires showing some causal connection between *explanandum* and *explanans*. Proofs establish necessary connections for Galileo by virtue of their structural characteristics in much the same sense as arguments are said to be valid. But explanations do more than that. The function of an explanation is to provide empirical knowledge. Thus, while a proof *qua* proof can exhibit logically necessary connections, thereby yielding certainty with respect to the connections in the proof, an explanation is supposed to be about the world. And, while Galileo was convinced of his ability to use geometry to advantage, he was not so sure about providing certainty with respect to empirical relations. Even for Galileo, geometric proof in the tradition of Archimeides produced only logical certainty.

McMullin relies on the ambiguity in our (and supposedly Galileo's) rendering of the Greek term "apodeixis" to slip by the ambivalence between proving and explaining. He (McMullin) notes that 'showing' has three senses in both Greek and English — to prove, to explain, or to teach — and that to quality as fully scientific knowledge, what one produces must accomplish all these goals. [McMullin (1978, p. 213)] But it seems clear that Galileo, while offering *proofs*, is not always convinced he has *explained*. Take, for example, his proof for the tides in Day 4 of the *Dialogue*. It has two parts. First, he offers a geometric proof. But he feels compelled to urge by way of *explanation* a terrestrial analogy and a set of secondary causes. More on this later.

Take another example. In *The Starry Messenger* he provides an explanation of the irregularity of the surface of the moon by first drawing an analogy with mountains on the earth and then proceeding to construct a geometric proofs. The proof is used to show how the analogy between mountains on the earth and mountains on the moon can be defended.

This is the model of Galilean explanation. By geometric proof alone nothing physical is shown. On this point McMullin agrees. For Galileo to provide an explanation required more: interpretation of the proof in physical terms.

At this point, let us pause to forestall at least one possible objection. First, aren't I jumping all over the place, just as I accused McMullin, from the *Dialogue* back to *The Starry Messenger*. Yes, but there is a difference. It seems perfectly reasonable to argue for a uniform concept of explanation as a more limited target than for the broader notion of science. To find some general concept of science in Galileo, McMullin develops a complicated balancing act between two competing aspects of Galileo's procedure without any obvious principle for selecting these as fundamental or essential to Galileo's views on explanation. But, if a clean and clear concept of explanation can be found across Galileo's work, independently of any claims about Galileo's view on "science", the systematic employment of that demonstrably constant procedure would argue for its priority in Galileo's thought. Such a procedure can be idenitied, as is shown below, and its systematicity argues for its methodological priority. Once this appreciation of Galileo's method is in hand, one *might* want to go on to argue that to the extent that Galileo had a conception of science it depended on his concept of explanation. But that sense of explanation has to be shown first — not assumed.

What, then, is this account of explanation? Essentially, as already noted, it involves the production of a deductive argument within a model accompanied by a commentary (see Pitt 1980, 1981). The commentary is designed to express the limitations, analogical character, etc., of the use of the model, as well as to indicate its scope.

The move here is analogous to, but not limited in the same ways as, the logical positivist conception of a language as a calculus plus an interpretation.[6] Geometry and geometric proof provide the model for the production of a demonstration which provides certainty. But what the demonstration provides certainty *about* requires an accompanying commentary. McMullin wants to argue that the certainty concerns the proof, and in this I agree. In order for the demonstration to provide certainty about the world, the commentary would have to provide an interpretation of the model in a formal sense. This requires that this commentary be derived from a unified, coherent world view. Once he endorses Copernicanism, Galileo has no such view. Copernicanism is at most a mathematical theory based on the assumption that the planets have a different arrangement than that proposed by the geocentric view. It has no accompanying theory of the physical forces needed to account for this arrangement.

Now, it is a fact that Galileo holds constant to his reliance on geometric proof; but he also varies his interpretations or commentaries, as I have been calling his accompanying discussions. Sometimes he suggests a one to one correspondence between physical notions and geometric points, as in *The Starry Messenger* and his account of mountains on the moon. Sometimes he argues the correlation with further commentary on the limits of the correspondence as in the discussion of tides in the *Dialogue*.

What can we conclude from noting these variations? Well, it depends on the commentary offered, but in general it reflects two things; first, Galileo's increasing sensitivity to the magnitude of the problem posed by adopting the Copernican model as a physical system and second, a rather honest reflection of Galileo's own sense of what it is possible to know.

I can't defend the first claim now, so I will leave it. But I can defend the view that Galileo had limited expectations. To begin, there is his view of how much we can know. At the end of Day 1 of the *Dialogue*, he offers the distinction between intensive and extensive knowledge.

Extensively, that is, with regard to the multitude of intelligibles, which are infinite, the human understanding is as nothing even if it understands a 1,000 propositions, for a 1,000 in relation to infinity is zero. But taking man's understanding *intensively*, in so far as this term devotes understanding some proposition perfectly, I say the human intellect does understand some of them perfectly, and thus, in these it has as much certainty as Nature itself has. Of such are the mathematical sciences alone; that is geometry and arithmetic, in which the Divine intellect indeed knows infinitely more propositions, since it knows all. But with regard to those few which the human intellect does understand, I believe that its knowledge equals the Divine in objective certainty, for here it succeeds in understanding necessity, beyond which there can be no greater success (p. 103).

So, in terms of the *amount* we can know, geometry gives us access, but the totality of it all is beyond us. Secondly, with respect to the commentaries on the geometric proofs, we should expect little by way of regularity in terms of their content. In the first place, as already noted, the demand for such regularity presupposes on Galileo's part some general world view which includes some theory about what will count as evidence and what role the model should play. There is no reason to suppose that Galileo had such a view. If anything, we can argue that whatever he thought about the restricted reliability of geometry, he believed our ability to ever achieve knowledge, i.e., certainty about the world, was limited by virtue of both our constitution and Nature herself. In *The Assayer*, he claims:

I might by many other examples make clear the bounty of nature in producing her effects by means which we would never think of if our senses and experiences did not teach us of them, though even those are sometimes insufficient to remedy our incapacity. Therefore, I should not be denied pardon if I cannot determine precisely the manner in which comets are produced, especially as I never boasted that I could, knowing that it may occur in some way far beynd our power to imagine (*Opere*, Vol. 6, p. 281).

And again later in the *Dialogue* he has Sagredo note:

It always seems to me extreme rashness on the part of some when they want to make human abilities the measure of what nature can do. On the contrary, there is not a single effect in nature, even the ideals that exist such that the most ingenious theorist can arrive at a complete understanding of it (p. 100).

Finally, in the *Discourses*:

No firm science can be given of heaviness, speed, and shape which are variable in infinitely many ways. Hence, to deal with such matters scientifically, it is necessary to abstract from them.

And the abstraction results in idealizing them in geometric form. Galileo then continues with the observation that we need to appeal to experience, which produces the commentary we thereby use, in order to demonstrate the relevance of our proof to the world:

We must find and demonstrate conclusions abstracted from the impediments, in order to make use of them in practice under limitations that experience will teach us (*Opere*, Vol. 8, p. 276).

So, Galileo's theory of explanation involves two features — a geometric proof and an accompanying commentary. Variations in the content of the commentary and the extent of the claims made for the proof can be accounted for in terms of Galileo's own limited expectations and his appreciation of the magnitude of the problem of offering an explanation in what we today would call a theoryless context.

The primary vehicle for providing explanations rests on the use of geometry. Let us consider this further. Galileo never rejects a non-geometric proof because it is not geometric. If he rejects a proposed proof, it is because it is based on faulty assumptions or inconsistent principles; witness his constant worrying of Simplio's arguments in the *Dialogue* on just these grounds. In those cases, where Galileo introduces geometric proofs in addition to whatever other proofs have been offered it is by way of clarifying the point in contention. Thus, early in the *Dialogue*, when discussing whether or not a body moving on an inclined plane must pass through infinite gradations of slowness before acquiring its final velocity, Salviati says in response to Sagredo's "I do not quite understand the question" with "I shall express it better by drawing a little sketch" (p. 23). And later, Salviati does not suggest replacing Gilbert's arguments concerning the lodestone with geometric proofs, rather the expresses a certain displeasure with the absence of geometry for reasons of rigor. Thus, he says,

What I might have wished for in Gilbert would be a little more of the mathematician, and especially a thorough grounding in geometry, a discipline which would have rendered him less rash about accepting as rigorous proofs those reasons which he puts forward as *verae causae* for the correct conclusions he himself had observed. His reasons candidly speaking, are not rigorous, and lack that force which must unquestionably be present in those adduced as necessary and eternal scientific conclusions (p. 406).

So it is the rigor that is missing, not the content.

Finally, in defense of his ill-fated theory of the tides, and, in particular, after attempting a general account of the effect of the diurnal rotation of the earth on the growth of the tides, he continues with the familiar, "we shall see whether the drawing of a little diagram will not shed some light on it" (p. 457). Now the diagram itself only sheds the light intended as its parts are identified with the previous discussion.

Galileo's reliance on geometry, therefore, is to be seen in the context of his uncertainty about providing physical explanations of the world. Geometry is his touchstone for certainty in a world of infinite variation and magnitude that he has been partially responsible for bringing to our attention. But as he shows — the certainty of the proof is not an explanation of the phenomena.

II. THE ROLE OF THE ARGUMENT ON THE TIDES
IN THE STRUCTURE OF THE DIALOGUE

The Dialogue Concerning the Two Chief World Systems is divided into four parts, each called a day. At first glance the structure of the argument seems to follow what Galileo claims he is about in his Preface. There he says he first intends to show that "all experiments practicable upon the earth are insufficient measures for proving its mobility". Then he will examine the celestial phenomena, "strengthening the Copernican hypothesis" (p. 6).

Finally, "I shall propose an ingenious speculation that the unsolved problem of the ocean tides might receive some light from assuming the motions of the earth". These divisions accord nicely with the arguments of the second, third and fourth Days. Well, what about Day One? Stillman Drake calls this Galileo's "metaphysical introduction". [Drake forthcoming] Drake claims Galileo tacked it on after the rest of the *Dialogue* was finished in 1629, "to give his book some elegance beyond dry sciences." So, if the major argument concerns setting the stage for the tides, two questions arise. (1) How does Galileo see Days Two and Three as performing that function, and (2) what about Day One?

Beginning with Day Two, Galileo argues first that no earthly experiments can solve the question of the motion of the earth either way. Then he proposes the merit of the Copernican view. And as a proof of Copernicus, so the story goes, he argues in Day Four, if we assume Copernicus is correct, then we can explain why there are tides.

Well, what is that explanation? If my early account of the structure of Galilean explanation is correct, his account of the tides is a classic example of that method. Galileo's explanation of the tides consists of two parts: (a) a geometric representation, (b) a commentary to accompany it. In this case he gives the commentary first and then follows it with the geometry. The commentary here involves an example of a barge carrying water in its hold. The barge moves at irregular speeds, first going fast and then slowing down. It is observed that the water moves forward in the hold of the barge when the barge slows down and rushes to the back when it speeds up. Galileo then says:

What the barge does with regard to the water it contains, and what the water does with respect to the barge containing it, is precisely the same as what the Mediterranean basin does with regard to the water contained within it, and what the water contained does with respect to the Mediterranean basin its container — ... The next thing is for us to prove that it is true, and in what manner it is true that the **Mediterranean and all other** sea basins move with a conspicuously uneven motion, even though nothing but regular and uniform motions may happen to be assigned to the globe itself (pp. 425—6).

So, two things — first there is the claim of a parallel between the water in the barge and the water in the Mediterranean. Second, he is going to *prove* (i.e. provide a demonstration) that the basins move irregularly, producing tides,˙even though only uniform motion can be claimed for the earth as a whole.

The proof shows how the two motions attributed to the earth, rotation on its axis and rotation around the sun when compounded one with the other produce an uneven motion in parts of the earth. Consider Galileo's own words:

From the composition of these two motions, each of them in itself uniform, I say that there results an uneven motion in the parts of the earth. In order for this to be understood more easily, I shall explain it by drawing a diagram.

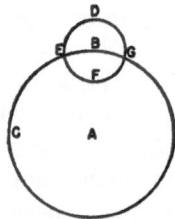

First I shall describe around the center A the circumference of the earth's orbit BC, on which the point B is taken: and around this as center, let us describe this smaller circle DEFG, representing the terrestrial globe. We shall suppose that its center B runs along the whole circumference of the orbit from west to east; that is, from B toward C. We shall further suppose the terrestrial globe to turn around its own center B from west to east, in the order of the points D, E, F, G, during a period of twenty-four hours. Now here we must carefully note that when a circle revolves around its own center, every part of it must move at different times with contrary motions. This is obvious consider-ing that when the part of the circumference around the point D is moving toward the left (toward E), the opposite parts, around F, go toward the right (toward G); so that when the point D gets to F, its motion will be contrary to what it was originally when it was at D. Moreover, in the same time that the point E descends, so the speak, toward F, G ascends toward D. Since this contrariety exists in the motion of the parts of the terrestrial surface when it is turning around its own center, it must happen that in coupling the diurnal motion with the annual, there results an absolute motion of the parts of the surface which is at one time very much accelerated and at another retarded by the same amount. This is evident from considering first the parts around D, whose absolute motion will be very swift, resulting from two motions made in the same direction; that is, toward the left. The first of these is part of the annual motion, common to all parts of the globe; the other is that of this same point D, carried also to the left by the diurnal whirling, so that in this case the diurnal motion increases and accelerates the annual motion (pp. 426—427).

So much for the tides.

That still leaves us with the purpose of Day One. The interesting thing here is that Drake may have been a bit too conservative. If Drake is correct, Galileo is forced to use a metaphysical theory — the one outlined in Day One to jazz-up his presentation. However, on my view, Day One is not *primarily* concerned with a metaphysical theory. It does invoke some metaphysics, specifically the view that geometry provides some mitigated access to the world and that geometric proof also provides certainty. But the discussions of Day One are best and most easily seen as Galileo's account of what consitutues the best way to provide an explanation.[7] Thus the major point of Day One is to establish the superiority of geometric proof over Aristotelian arguments from first principles. The discussion begins with Galileo attacking Aristotle's proof that the world is perfect because it has all three dimensions — length, breadth and depth. Note, Galileo does not dispute the conclusion, that the world is perfect — rather "I much wish Aristotle had proved to me by rigorous deductions that simple length constitutes the dimension which we call a line, etc." (p. 9—10). But, there is more, for Galileo is not merely calling for a demonstration, he wants a demonstration of Aristotle's first principles — thus he goes on to comment:

To tell you the truth, I do not feel impelled by all these reasons to grant any more than this, that whatever has a beginning, middle, and end may and ought to be called perfect. I feel no compulsion to grant that the number three is a perfect number, nor that it has a faculty of conferring perfection upon its possessors. I do not even understand, let alone believe, that with respect to legs, for example, the number three is more perfect than four or two: neither do I conceive the number four to be any imperfection in the elements, nor that they would be more perfect if they were three. Therefore it would have been better for him to leave these subtleties to the rhetoricians, and to prove his point by rigorous demonstrations such as are suitable to make in the demonstrative sciences (p. 11).

Now this is Galileo's *modus operandi* for the rest of Day One. In each case, the first move is to show that Aristotle could have done better. The second move is to subtly introduce physical objects into the geometric demonstration — so that in the argument on the acceleration of bodies on an inclined plane we quickly find ourselves discussing the attributes of real balls moving on real inclined planes impeded by terrestrial imperfection, having initially started with a simple geometric diagram of a triangle. The same strategy is found in Day Two, where Galileo is arguing over whether two material — i.e. physical, spheres touch each other at more than one point, the move is bold — the argument follows from Galileo's metaphysical point (pace Drake) at the end of Day One about the priviledged access to the world which geometry provides:

Then whenever you apply a material sphere to a material plane in the concrete, you apply a sphere which is not perfect to a plan which is not perfect, and you say that these do not touch each other in one point. But I tell you that even in the abstract, an immaterial sphere which is not a perfect sphere can touch an immaterial plane which is not perfectly flat in not one point, but over a part of its surface. so that what happens in the concrete up to his point happens the same way in the abstract. It would be novel indeed if computations and ratios made in abstract numbers should not thereafter correspond to concrete gold and silver coins and merchandise. Do you know what does happen, Simplicio? Just as the computer who wants his calculations to deal with sugar, silk, and wool must discount the boxes, bales, and other packings, so the mathematical scientist (*filosofor geometra*), when he wants to recognize in the concrete the effects which he has proved in the abstract, must deduct the material hindrances, and if he is able to do so, I assure you that things are in no less agreement than arithmetical computations. The errors, then, lie not in the abstractness or concreteness, not in geometry or physics, but in a calculator who does not know how to make a true accounting (p. 207).

The conclusion seems obvious. The explanatory force of the argument is carried first by the geometric proof and then by the accompanying commentary which *more or less* ties the proof to empirical phenomena.

Now — if we return to the tides — I believe that Galileo is showing here that a certain kind of demonstration provides results not previously obtainable. The explanatory force of the proof lies in the success of the commentary in which a correspondence is pointed to between certain aspects of the proof and key features of the world. Or, to phrase it differently, the geometric part of the proof is an explanatory sketch. If the sketch can be filled in by appeal to empirical correspondences, that argues for the proof as an explanation. No forces are needed — no unified theory of physics is required — just a one to one correspondence between empirical observations and geometric features. Thus, even if the particulars vary, the indisputability of the geometric proof remains. It is for this reason that Galileo did not feel troubled by the fact that only one high and one low tide a day followed from the geometry. He at least had *that*, and could fill in the rest by reference to secondary causes.

III. EXPLANATION AS JUSTIFICATION

At the beginning of the discussion Galileo was characterized as an explanationist — that is, as one whose approach to justification consists in appealing to explanations rather than the weight of evidence or a confirmation function. The question remains as to the adequacy of this method of justifying one's conclusions. Specifically, is it legitimate to claim one is justified in maintaining a given proposal, P, on the grounds that by maintaining P one is in *the best position* to explain either what has not hitherto been explained or what has been explained less well.

The problem is that a Galilean explanation does not make it clear that one is in such a position. For, as we have seen, a Galilean explanation amounts to only a physical, viz. terrestrial, interpretation of a set of geometrical propositions and an analogical extension of that interpretation to another set of physical phenomena. What one wants to know is whether the conclusion one draws from the extended inference is true. On Galileo's account there is no guarantee that the inference results in a true conclusion.

Galileo, aware of this problem, was not insensitive to it. In fact, it seems to motivate most of his strategy. His tactic was to take observationally justified terrestrial examples and extend them by parity of reasoning to the celestial domain. This approach flounders because it appears to beg the question; namely, is it correct to assume that the heavens obey the same laws that operate on the earth? Obviously Galileo thought he was making some kind of headway on this issue with his telescopic observations of the phases of Venus and the moons of Jupiter. But that too seems to beg the question. Those observations were terrestrially based using an instrument constructed in accordance with the laws governing terrestrial phenomena.

Despite these flaws, Galileo was convinced of the reasonableness of his method. The basis of his conviction was what I call his geometrical realism. As we saw in a number of quotes drawn from the *Dialogue*, Galileo believed that God organized the universe in accordance with the principles of geometry and that to the extent that we provide geometric accounts of the movements of the planets, etc. we could have knowledge as certain as God's own. Galileo's realism, then, comes in two dimensions: ontological and epistemological. He believed the universe existed as geometrically determined and he believed we could come to have knowledge of that determination. Furthermore, since the use of geometry as an appropriate tool for describing the heavens was not disputed and since geometry was equally well accepted for terrestrial purposes, Galileo did in fact have the foundation for a unified general theory. By appeal to the traditions of both Ptolemy and Archimedes he also had authority behind him. Geometry was the primary tool of the astronomers and mechanics was becoming increasingly mathematized.

The question remains as to whether geometrical realism in turn justifies Galileo's method of justifying claims about the heavens by appeal to the explanatory power of extended inferences drawn from terrestrial examples. This, in turn, raises the broader question of the role of realism in scientific theorizing in general. And while this latter is an issue of enormous concern, it is not clear that it can be resolved. The battle lines are drawn between scientific realists and non-realists in much the same way as we used to distinguish realists from instrumentalists.[8] The scientific realist believes that there is a determinate structure to the universe which is discoverable by men, the non-realists minimally denies that the real structure and make-up of the universe can be known. Varieties of both realists and non-realists are distinguished from each other and one another by the degree of commitment they exhibit to claims about the reality of the universe and the kinds and degrees of knowability.

While it is not clear that we can resolve the scientific realist/non-realist issue, we can say something about Galileo's geometrical realism and the role it plays for him. From our perspective, Galileo's appeal to the role of God in all this seems an inadequate defense of geometry as a universal basis for a general unified theory of physics. On the other hand, the degree of certainty that mathematics permits is and was unexcelled. It does not, therefore, seem reasonable to dismiss that commitment because of extraneous claims of divine origin. Furthermore, if we make an effort to construct an historical overview we will find Galileo in good company when it comes to having a commitment to the ontological primacy of mathematics. In his belief in the universality of geometry, we can see a foreshadowing of some modern views in which mathematical entities are proposed as the ultimate constituents of the universe. But that is also an old view, Pythagorean in origin, carried on through various versions of Platonism. This is all fine and good, but if we try to make a case for the reasonableness of Galileo's geometric *realism* on these grounds, it will also get us embroiled in some of the kinds of disputes which have often taken scholars away from trying to understand Galileo. What seems to be important about Galileo's strategy and his use of geometry is not that he can be seen as a prophet of the new physics or as another figure in a long tradition. It is rather that there were good reasons for his use of geometry: it enjoyed wide spread acceptance, it provided clear demonstrations, it added rigor and gave a sense of necessity in the proof. Convinced of this, it then seems appropriate for Galileo to use Day One of *Dialogue* to show how geometry can be used to do more than provide proofs, i.e., it can be used to explain actual events. And because geometry has long been used to discuss matters both terrestrial and celestial, it is not question begging to show how on the assumption of certain celestial motions, i.e. the planets around the sun, the tides can be explained. But because of the magnitude of the required assumption of the motion of the earth around a stationary sun, it is not clear that Galileo had the *best* explanation of the tides. For, one thing, we don't know how to compare explanations which proceed according to different criteria of success, i.e. Aristotelean and Galilean. Therefore, we can only conclude with respect to what Galileo thought, that the price involved in making this assumption was not too high; in other words, that his explanation of the tides was justified and that his strategy in presenting his proof was rational.

NOTES

Earlier versions of this paper were presented to the Canadian Society for History and Philosophy of Science (Montreal, 1981) and to the Center for the Study of Science in Society at Virginia Tech in February 1984. I wish to thank Stillman Drake, William Shea, Larry Laudan and Richard Burian for helpful comments. I hope they will forgive my not taking all of them to heart; for if I had, there probably would be nothing left.

[1] For an analysis of the relation between Galileo's account and that of the other major tideal theories of his day, see Shea (1972). Roger Ariew's (1984) looks at earlier Lunar theories and their merits. For a detailed account of Galileo's development of the theory of the tides see Drake (1970) and (1978).

[2] Feyerabend (1975) argues to the contrary that there is nothing questionable about Galileo's approach. On Feyerabend's view rhetorical success is the objective here, and since Galileo's strategy is aimed in that direction, there is nothing unreasonable about it.

Feyerabend may be right about the rhetorical value of the *Dialogue*, but more seems to be going on in that work than a mere public display of cleverness.

[3] "Explanationism" is a term introduced by Keith Lehrer in his (1970) in the following way:

> A belief is justified by its explanatory role in a system of beliefs. Some beliefs are justified because of what they explain, and other beliefs because they are explained, but every belief that is justified is so either because of what it explains or what explains it. These doctrines formulate a theory of justification which I shall call explanationism.

Gilbert Harman has a variant of that theory in which the justification is located in the *inference* to the *best* explanation. See his (1973). In her (1983) Nancy Cartwright argues against explanationism. For an examination of Wilfrid Sellars' fully developed philosophy of science in which explanationism plays a crucial role, see Pitt (1981).

[4] For an excellent account of Galileo's use of reasoning *ex suppositione* see Wallace (1978).

[5] In fact, McMullin *begins* with the *Discourses* (which is Galileo's last work) and uses Galileo's method there as paradigmatically representative of his earlier views. This is somewhat like observing that in Beethoven's final symphony there is a major choral component and then trying to find out where the choral material belongs in the earlier eight symphonies.

[6] See Pitt (1981).

[7] This is essentially the same strategy Galileo follows in Day One of the *Discourses*. He first establishes the adequacy of his justificatory procedures and then moves to his scientific discussions.

[8] For a good presentation of arguments representing the status of the debate between realists and non-realists see Leplin (1984).

REFERENCES

Ariew, Roger. (1984). "Galileo's Lunar Observations in the Context of Medieval Lunar Theory". *Studies in History and Philosophy of Science*, Vol. 15.

Cartwright, Nancy. (1983). *How the Laws of Physics Lie*. Oxford: Clarendon Press.

Drake, Stillman. (1970). *Galileo Studies*. Ann Arbor: University of Michigan Press.

Drake, Stillman. (1978). *Galileo at Work*. Chicago: University of Chicago Press.

Drake, Stillman. (Forthcoming). "Re-examing Galileo's *Dialogue*."

Galileo, G. (1968). *Le Opere*, Vols. 1—11. Firenze: G. Barbara.

Harman, Gilbert. (1970). *Thought*. Princeton University Press.

Koyré, Alexandre (1939). Études *Galiléennes*. Reprint, Paris: Hermann.

Lehrer, Keith. (1970). "Justification, Explanation, and Induction". In Swain, M., ed. *Induction, Acceptance, and Rational Belief*. Dordrecht: D. Reidel.

Leplin, Jarrell. (1984), ed. *Scientific Realism*. Berkeley: University of California Press.

McMullin, Ernan. (1978). "The Conception of Science in Galileo's Work". In R. E. Butts and J. C. Pitt, eds. *New Perspectives on Galileo*. Dordrecht: D. Reidel. Pp. 209—258.

Pitt, Joseph C. (1980). "Hempel Versus Sellars on Explanation". *Dialectica* **34**: 95—120.

Pitt, Joseph C. (1981). *Pictures, Images and Conceptual Change*. Dordrecht: D. Reidel.

Pitt, Joseph C. (1983). "The Untrodden Road: Rationality and Galileo's Theory of the Tides". *Nature and System* **4**: 87—99.

Shea, William. (1972). *Galileo's Intellectual Revolution; Middle Period, 1610—1932*. New York: Scientific History Publications.

Wallace, William. (1978). "Galileo Galilei and the *Doctores Pariseienses*". In R. E. Butts and J. C. Pitt, eds. *New Perspectives on Galileo*. Dordrecht: D. Reidel. Pp. 87—138.

Woolhouse, R. R. (1983). *Locke*. Minneapolis: University of Minnesota Press.

WILLIAM R. SHEA

THE QUEST FOR SCIENTIFIC RATIONALITY:
SOME HISTORICAL CONSIDERATIONS

I. THE PROBLEM OF RATIONAL CHANGE

These are indeed exciting times for the philosophy of science. With the
decline of logical empiricism practitioners have spread themselves over
a wide range of alternatives, resuscitating, in the process, philosophies
that the positivists had condemned to the netherworld of obscurantism
and irrationality. Hence the objection that the new look is merely a
facelift or, more seriously, the charge that rationality has been abandoned for pre-rational, a-rational or irrational modes of thought. But
while it may be true that philosophy of science can no longer be
described and justified as a body of established knowledge, the death-
knell of positivism need not toll for rationality as such. Where one
brand of rationality failed, another may thrive.

Two of the main positions that flourish at the present time are
scientific realism and relativism or radical contextualism. Both reflect a
shift of interest from scientific data and how it is known to scientific
theory and how it is constructed. Both take it for granted that a
scientific theory is necessarily underdetermined by the data it describes,
but from this springboard they jump in different directions. Whereas
the realist couches his analysis of scientific theory in terms of metaphysics, the relativist questions the possibility of making such an
analysis in philosophical terms at all, and maintains that theories are
only intelligible in the social and cultural context in which they are
born. The radical contextualist is therefore inclined to consider that no
theory is irrational so long as it is espoused by some group or used to
further the aims of a given society. To a realist, this is a subversion of
rationality itself since, with a bit of ingenuity and goodwill, a reason can
be found for anything.

This debate is unlikely to be closed in the immediate future and all I
wish to offer in this paper are some reflections on the genesis of our
present predicament. I shall consider some epistemological aspects of
the problem — in particular, the relations between conceptualization
and experimentation — before examining in some detail one instance of

155

J. C. Pitt and M. Pera (eds.), Rational Changes in Science, 155—176.
© 1987 by D. Reidel Publishing Company.

scientific change, the shift from the Cartesian theory of vortex motion to the Newtonian theory of gravitation.

II. THE LANGUAGE OF DESCRIPTION

I begin with one of Max Born's famous anecdotes.[1]

A lady once asked a professor at a dinner-party: "Can you tell me in a few simple words what the theory of relativity is all about?" The professor replied,

With pleasure, but let me tell you a little story first. I was going for a walk with a French friend and we both got thirsty. By and by we came to a farm and I said: "Let's buy a glass of milk here". "What's milk?" "Oh, you don't know what milk is? It's the white liquid that —". "What is white?" "White? You don't know what that is either? Well, the swan —" "What's a swan?" "Swan, the big bird with the bent neck". "What is bent?" "Bent? Good heavens, don't you know that! Here, look at my arm: when I put it so, it's bent!" "Oh, that's bent, is it. Thank you, now I know what milk is!"

The story does not tell us whether the lady wanted to hear about the theory of relativity after that, but the point that Born was trying to make is clear. It's the one that Quintilian made what he wrote "*paene omne quod dicimus metaphora est*" If we are to communicate at all we must do so in words, and words are not only themselves sensible but their initial meaning is rooted in sensible experience. By a gradual series of transformation this meaning evolves until the primary reference to sensible objects is submerged or forgotten, and new abstract meanings emerge. But the new meanings can only be communicated if a speaker knows how to choose the words that will effectively register with his audience. If he feels that his hearers do not have the tools to grasp his meaning he will experience frustration and pass over the matter with a joke. The professor who was asked at a dinner party to explain relativity to a lady innocent of physics and mathematics obviously felt that the task was impossible. Had he been asked why, he would probably have pointed out that her interest in the theory was quite different from his own ("She wants to know if we can travel backwards in time"), that it took him years to learn what he knows, and that the process of learning cannot be telescoped. He might have added that popular circumlocutions really confuse the issue and clog the process of learning.

Anyone who has had to lecture to general audiences will feel some sympathy for the professor. Still the first natural philosophers were under the necessity of remaining silent or of communicating with ordinary people in ordinary language. They had to bring about the transformations of meaning that change the reference of words from the sensible to the intelligible, and they had to do this without the tools that would enable them to explain to themselves or to others precisely what they were doing. If the philologist can work backwards from our meaning of a word through a series of other meanings to the initial meaning of the root, there must have existed a series of discoveries of new meanings. As long as such changes were mere developments of existing viewpoints, the new meanings could be imparted by employing old words outside their habitual context. But when discoveries ushered in new viewpoints, a more elaborate procedure became necessary. This is why Plato, for instance, introduces in his dialogues myths to convey insights and evaluations that would seem strange and novel.

The logical positivists were conscious of this sensible origin of our most abstract words but in their eagerness to find an epistemological rock-bottom they assumed that genuinely scientific terms could be anchored to their sensible roots in an unproblematic way. In doing this, they took for granted that scientific notions were derived from commonsense ones by a process of expansion, and they excluded the possibility that they might have resulted from a radical transformation akin to the one that Plato strove to express. Their quest was one-sided and the neo-positivists themselves subsequently cast their nets into deeper and deeper waters. Some even came to realize that their original motto, "What My Net Doesn't Catch Isn't Fish", made real anglers smile. Art historians — E. H. Gombrich most prominently — have insisted that no painter can look at, conceive, or represent nature directly "as it is"; he has to work within some established framework of defined images, symbols and schemata which he may then adapt or modify to his own purposes. In Popper's arresting metaphor, knowledge, scientific or otherwise, "is like a building erected on piles",[2] or, as W. V. Quine insists, there is no "cosmic exile" who can view the world in an altogether detached, objective point-of-view-less way.[3] Reality, any reality, "our reality", is a mental construct built up by means of a conceptual scheme that invariably and inevitably makes essential use of organizing principles. If we do not have them we cannot begin to learn. This is why some philosophers, such as Descartes, believed that they were ultimately innate. Without them there is no cognitively meaningful experience. They are *a priori*, if only in the relative sense that they cannot be read off from the world about us by some concept-free observational process of mere inspection. Indeed with observation and inspection it is already too late to reach them since, as Kant remarked, raw experience uncooked by conceptualization is not possible. Whether we ascribe our ability to rationally structure experience to an innate quality or to an evolutionary development that lasted millenia, the machinery is indispensable.

All this sounds pretty basic but we must remember that positivists looked at sensations rather than theories. Yet even here, their confidence is hard to understand. Trained in schools where Greek was still taught, they must have been aware of the problem that translators face with apparently innocuous words such as colors. Euripides says in *Iphigenia* that blood made the side of the altar *ksanthon*. The word would therefore appear to mean *red*, and Latin translators rendered it by *ruber*, the Latin for red. Aeschylus in the *Persians*, however, calls the leaves of the olive tree *ksanthon*. Thus it would seem to indicate *green*. But *ksanthon* is also used to describe the color of honey and hence included yellow also. Other colors receive an equally striking treatment at the hands of the Greeks. *Ockron* is the common word for the color of the skin and is usually rendered by *pale*. In a medical treatise, however, the question, When is fever diagnosed? is answered with, When the patient turns *ockron*. Hippocrates even says, on one occasion, that *ockron* is the color of fire. To add to our embarrassment, yet another writer speaks of the color of a toad as *ockron*. Does it also denote green or perhaps brown?

Glaukon offers a third example. The color of the moon is *glaukon* and fiery eyes are called by that name. It seems therefore to be a glowing red. But Plato explains *glaukon* as the mixture of dark blue and white, and this points to its being sky-blue or blue-grey, *caesius*, as the Latins rendered it. Again Euripides speaks of a vestibule decorated with leaves as *glaukon*, hence as related to the color of vegetation. Many translators seek a way out by rendering *glaukon* as *brilliant* or *shiny*. For instance, in the Loeb Classical Library, for "glaukopis Athene" we read "flashing-eyed Athene". But this solution is itself highly problematic since the eye-disease that we call by its Greek name *glaucoma* does not give the eye a shiny but a turbid appearance. A more radical solution would be to postulate that the Greeks were color-blind. But this is a counsel of despair. What is needed is not an eye doctor but a linguist or a cultural anthropologist!

There are well ascertained cases of people having suffered damage to the speech center in their brain. In one such case, the therapist asked the injured person to order a bundle of colored threads. The patient had no problem identifying individual colors, for instance, he never erred in choosing the color of strawberries or cornflowers. But when asked to single out all the red threads, he was at a loss. He saw each color as a class unto itself. He was unable to comprehend under one word "red" the shades of color from pink to crimson. The damage he had sustained had robbed him of the notion of redness. The world no longer came to him in the simple seven colors that we habitually distinguish in the rainbow. He had no longer the compulsion or even the ability to divide the continuous spectrum of colors at the traditional points.

Since red and blue are common and convenient labels for a certain width of optical sensations, we would be surprised if some communities were to cluster their sensations in a different way. It seems an obvious enough fact that the color of the rockrose and the color of blood have a striking similarity that we express by naming them *red*. But a different civilization may be struck by the fact that the color of a rockrose has something in common with the color of a birch-leaf, namely the same *brightness*. As a result, they might fashion a concept to include pink and light green and never devise our categories of red and green.

Two last examples, this time from the Romans who are closer to us and cannot be suspected of collective color-blindness. They concern purple and blue. Anyone assuming that the English for *purpureus* is *purple*, would find himself in deep linguistic trouble when he discovered that the poet Horace calls a swan *purpureus*. Indeed, the Latins gave shining black as well as shining white objects the name *purple*. Or consider *caeruleus* (deep blue) used to describe the sky, the sea and the eyes of Germans, but also the hair of Indians, the night and death!

It is clear from these illustrations that a variety of different conceptual frameworks can be superimposed with equal validity upon the order of "colored things". The "colors" that we encounter do not by themselves determine how we are to think of them independently of the apparatus of thought that we bring to the transaction. There are, as we have seen, many possibilities but none are independent of the conceptual framework of the experiencing agent. The framework fundamentally conditions our perception. Seeing is always seeing-that: we never experience seeing something without "seeing something as such and such", and so seeing that it is such-and-such. We not only see upturned lips and furrowed brows but cheerful faces and worried looks.

This may seem like a great ado to make Kant's point about the essential inseparability of perception from conceptualizing thought. But what caused such disarray among some philosophers of science was the obvious problem that followed from this recovery of an elementary insight. If science is a conceptual schema among others, how do we safeguard its status as objective knowledge? How do we justify giving up a conceptual scheme in favor of another if there are no crucial experiments that sit in judgment on theories? The situation appeared fraught with difficulties but only because of a prior dogmatic belief in the availability of a tribunal called rigorous experimentation. Given that science is a conceptual network that is mind dependent, we can still grapple with the way it works. Why give up a theory in favor of another? As before the demise of positivism, the issue is one of cost-benefit evaluation. What would be gained and what would be lost by rejecting, for instance, the geocentric astronomy of Ptolemy for the heliocentric system of Copernicus? This question has received lavish attention from historians of science, and it is clear that at no time during the gradual acceptance of the Copernican system in the seventeenth century an astronomer would have been considered irrational for maintaining that the earth is at rest, and insisting that the hypothesis of Tycho Brahe would win out in the long run. Such astronomers were made fun of by Galileo in the first half of the century and by Fontenelle in the second, but for a long time choice hinged on preferences that were largely a matter of taste. The Copernican system was embraced by men who were willing to sacrifice the evidence of their earthly senses for a celestial harmony that was beyond the ken of their best telescopes. For Galileo, mathematics sharpened the wits of men and enabled them to penetrate beyond the veil of the senses. In a passage in his *Dialogue* he praises those who have

through sheer force of intellect done such violence to their own senses as to prefer what reason told them over that which sensible experience plainly showed them to the contrary ... I repeat, there is no limit to my astonishment when I reflect that Aristarchus and Copernicus were able to make reason so conquer sense that, in defiance of the latter, the former became mistress of their belief.[4]

Kuhn and Feyerabend have made this passage familiar to thousands of readers, but it is interesting to note that over a hundred years before the recent wave of radical contextualism, William Whewell had drawn attention to several of the more arresting features of scientific change in an essay entitled "Of the Transformation of Hypotheses in the History of Science".[5] I shall examine the nature of change in the light of Whewell's arresting study.

III. THE VORTEX THEORY OF PLANETARY MOTION

The first point Whewell stressed is the familiar one that is difficult for any one person to do justice at the same time to two conflicting theories. This can be seen in the clash within the Copernican school between the adherents of the Cartesian hypothesis of vortices and the supporters of the Newtonian doctrine of gravitation. The issue cannot have been trivial and the evidence cannot have been overwhelming since Fontenelle, the Secretary of the Académie des Sciences, died an impenitent Cartesian seventy years after the publication of the *Principia*! Yet since their complete triumph, the Newtonians have been unwilling to admit the slightest virtue in the doctrine of vortices. "It cannot but seem strange to a calm observer of such changes", observes Whewell, "that in a matter which depends upon mathematical proofs, the whole body of the mathematical world should pass over, as in this and similar cases they seem to have done, from an opinion confidently held, to its opposite".[6] Prejudice and education, fashion and the youthful desire for novelty appear to offer some explanation when we see, for example, Johann Bernoulli continuing a Cartesian to the last while his son, Daniel, is a Newtonian from the first. But this is only a partial explanation and it should not be made to hide a more important feature of change. Contrary to the dramatic paradigm shifts that are now alleged to be the prototypes of scientific change, Whewell noted that when a prevalent theory is replaced by a different, or even by an opposite one, the change is not instantaneous but is affected by a transformation, or a series of transformations, of the earlier hypothesis that bring it nearer and nearer to the second. In this way, the defenders of the old doctrine are able to go on as if still asserting their first opinion while continuing to press their points of advantage if they have any. The new explanations are tailored to fit the original hypothesis in such a way that a verbal consistency is maintained while the first hypothesis gradually merges into the new one. An illustration of this process, which Whewell did not mention, is the work of the Jesuit astronomer Giovanni Battista Riccioli whose monumental *Almagestum Novum* of 1651 was the most important contribution to mathematical astronomy between Kepler and Newton. Riccioli held the geostatic system of Tycho Brahe but he adopted, and adapted to his system, the major achievements of his great Copernican predecessors Kepler and Galileo. The result, however impressive, lost its appeal after the victory of Newton, but the skill with which Riccioli combined both the new observations and the new mathematical techniques with Tycho's main hypothesis is a perfect illustration of what Whewell had in mind. The case study that Whewell himself develops is the struggle between the Cartesians and the Newtonians.

In the *Principles of Philosophy* and his posthumous *Le Monde*, Descartes explained the motions of the planets by supposing that they were carried around the sun by a kind of whirlpool of fluid matter in which they were immersed. He accounted for the rotation of satellites around the planets by similar subordinate whirlpools that were moved along by the primary vortex.[7] For us, this is the main feature of Descartes' system, but it must be borne in mind that for his contemporaries another part of his system was considered as important, namely the explanation that it was believed to afford of falling bodies. According to Descartes, terrestrial gravity or weight resulted from the motion of the vortex of celestial matter around the earth's axis. Since celestial matter had more centrifugal force than terrestrial bodies, if an object, say a stone, was released above the surface of the earth, it would be pushed downwards by the celestial matter rising to take its place.[8]

The Cartesian vortices fulfilled two functions: (a) they carried bodies in their stream, as straws are carried round by a whirlpool and, (b) they pressed bodies to the centre by the centrifugal effort of the whirling matter. These must be considered separately because they were modified separately. The first effect, the dragging forces of the vortex, was soon acknowledged to be at variance with Descartes' overriding principle of inertia. Descartes had rejected the Keplerian notion of an inertness in matter that constantly retarded the motion of bodies, and he had maintained that a body moving in a straight line would continue to do so unless an external force applied to it. This made it plain (if not to Descartes, at least to his followers) that a planet perpetually dragged onwards in its orbit by a fluid moving faster than itself would be constantly accelerated. This would render unintelligible the speeding up and slowing down that was observed when they approached or receded from the sun. Cartesians, then, did not take their stand on this first effect: they discreetly discarded it as incompatible with *a more fundamental* tenet of Cartesianism. This would, in effect, bring them closer to Newton who had enshrined inertia in the first of his famous three laws, but we must remember that for continental scientists this law was Descartes' before it was Newton's! Cartesians therefore focused their attention on the assumption that a vortex would produce a tendency of bodies to the centre, and they developed various mathematical strategies to construct vortices with the centrifugal force required to account for the falling of heavy bodies on earth.

There was one objection, however, that was formidable enough to give rise to a debate that occupied several of the Wednesday meetings of the Académie Royale des Sciences between 7 August and 20 November 1669. If terrestrial gravity, it was urged, arose from the centrifugal force of a vortex that revolves about the earth's axis, terrestrial gravity ought to act in planes perpendicular to the earth's axis, instead of tending to the earth's centre. Huygens, who took part in this debate, applied his knowledge of the laws of centrifugal force to calculate the mechanical action of the subtle matter and, in the process, converted Descartes' qualitative hypothesis into a model for mathematical physics. Descartes had failed to find a convincing experiment to vindicate his claim that the subtle matter, in order to produce the effect of gravity, must have been moving faster than the heavy body. Huygens devised just such an experiment. He placed a small globe in a cylindrical vessel filled with water and constrained it to move between three strings that just touched its circumference. The vessel was rotated and suddenly brought to rest: the water continued to rotate, and the globe, constrained by the strings, moved to the centre. Huygens claimed that this experiment showed that the speed of the subtle matter could be the cause of the motion of heavy bodies on the earth. The circulating subtle matter imagined by Descartes was vindicated but this left the difficulty of the path of fall: if bodies behaved like the globe, the circular motion of the subtle matter about the axis of the earth would only make them fall toward the axis and not towards the earth's centre. Huygens met this problem by formulating what was, for him, an ancillary hypothesis, and, for his opponents, an arbitrary assumption. He supposed that particles of the fluid matter can move in every possible direction within the spherical space which includes terrestrial objects, but that the greater part of these motions soon become circular and hence produces a tendency towards the earth's centre. He anticipated the objection that, since some circular motions would be clockwise while others would be anticlockwise, they would oppose and hinder each other by stating that the smallness and mobility of the parts of the fluid matter enabled it to sustain these different agitations, on the analogy of boiling water. Huygens explained that heavy bodies are not carried along in circles by the subtle matter, because they are pushed so quickly and in so many directions that there is not enough time for them to acquire any sensible horizontal motion. Hence they have no centrifugal force relative to the earth and they are displaced by the subtle matter towards the centre of motion, namely the centre of the earth.[9] This solution, implicitly identifying a vortex of some kind with a central force, made the vortex theory applicable wherever central forces existed but then, as Whewell points out, "in return, it deprived the image of vortex of all that clearness and simplicity which had been its first great recommendation".[10]

A related difficulty had also to be faced. According to this explanation of gravity, since the tendency of bodies to the earth's centre arose from the greater centrifugal force of the whirling matter which pushed them onward as water pushes a light body upward, heavy bodies ought to tend more strongly to the centre in proportion as they are less dense. The rarest bodies should be the heaviest, contrary to what we find. Descartes had coped with the objection in his *Principia* by postulating that terrestrial bodies consisted mainly of particles of the "third element": the more terrestrial a body the less its content of celestial matter (the "second element") and hence the greater its density and heaviness. This solution, however, supposes that celestial matter is entirely different from ordinary matter. It was soon realized that this was an obstacle to the application of mechanical principles developed for terrestrial objects. If the new mechanics was to be extended to the celestial matter said to produce the gravity of heavy bodies by its whirling, it must be very dense, in the ordinary sense of the term dense. Huygens replied that there were two meanings of density or rarity: some fluids might be rare by having their particles wide apart, others by having their particles very small although close and actually touching.[11] Joseph Saurin, who presented a memoir to the Académie Royale des Sciences in 1709, discussed Huygens view and refined it. According to him, two main factors determine the force of a fluid: the *density* or fraction of the total volume occupied by the particles of celestial matter, and the *ease of separation*, roughly the reciprocal of viscosity. Saurin's development of this idea reveals how a theory is gradually modified or even transmogrified. Saurin first showed that the force depended on the density, and that a fluid twice as dense as another would exert twice the force of the second. He then considered that size might be the most important factor affecting the ease of separation. This would have entailed that the force of the celestial matter could have been decreased by reducing the size of the particles. But the idea was discarded when Saurin discovered, after Newton, that the size of the particles did not alter the resistance offered by a fluid to solid bodies moving through it. At this point, another consideration obtruded itself and shifted the problem to a different level of discourse. Saurin was reminded that solid bodies, in Descartes' view, contain particles of celestial matter between their larger terrestrial parts. But if the particles of celestial matter are small enough to pass unhindered through the closest bodies, their force of impact must be vastly reduced. Saurin had no way of estimating the portion of celestial matter in ordinary bodies, but he ventured to guess that it might be as little as a hundred-thousandth.[12]

Assuming, for the moment, that in some way or other a centripetal force arises from the centrifugal force of the vortex, let us consider the important step Huygens had taken in his *Discours sur la cause de la pesanteur* when he applied mathematics to the hypothesis.[13] On the assumption that the weight of a body is equal to the force that an equal volume of subtle matter has to recede from the centre, and armed with his theorems on centrifugal force, Huygens calculated that the speed of the subtle matter in the vortex was seventeen times that of the earth's rotation at a point on the equator. The great speed of the celestial matter accounted from the constant acceleration of falling bodies. But if the celestial matter moved with a speed seventeen times that of the earth (and hence went round the earth in 85 minutes) would it not sweep all terrestrial objects with it? Huygens had no fear of this since the particles of the fluid celestial matter move in all directions, and therefore neutralize each others's actions, so far as lateral motion is concerned. Huygens was so pleased with his successful quantization of the vortex theory that he offered, as a further confirmation of its worth, Newton's proof that the centripetal force is equal to the centrifugal at the distance of the moon, and hence, that this force is less than the centripetal force at the earth's surface in the inverse proportion of the square of the distances.

Another interesting feature of Huygens' vortex theory of gravity was its ability to explain why the period of oscillation of a pendulum is slower at the equator than at the poles, a fact that Richer had ascertained with his famous pendulum experiment in 1672. Since Huygens had calculated that if the earth turned seventeen times more quickly the centrifugal force at the equator would equal the weight of the body, it followed that the diurnal motion must take away a fraction of the weight. As the centrifugal force was proportional to the square of the speed, this fraction was $1/(17)^2$ or $1/289$, and Huygens showed that the observed variation in the period of the seconds-pendulum oscillating at Cayenne and at Paris agreed with this. Huygens carried his achievement a step further by noting that if the earth were a perfect sphere, the diurnal motion of the earth would deviate a plumb line from the vertical by as much as 5' 54". Since no such deviation was detected, Huygens concluded that the earth was an oblate spheroid. This important discovery was arrived at independently by Newton along a somewhat different route, and this has tended to obscure the fact that Huygens made it by applying his theory of centrifugal force to the Cartesian hypothesis of the circulation of subtle matter.

IV. ADAPTING FOR SURVIVAL

When Newton died in 1727, the year following the publication of the third edition of his *Principia Mathematica*, natural philosophers on the Continent still felt that his objections to the theory of vortices could be countered but they expended increasingly more time in answering them. Johann Bernoulli, who won the prize for the 1730 Essay Competition of the Académie Royale des Sciences on the causes of the elliptical orbits and the rotation of the apsides, vigorously questioned Newton's mathematics, and redid, in the light of his corrections, Newton's calculations of the velocities of the layers in cylindrical and spherical vortices that were alleged by Newtonians to crush Descartes. The results of Bernoulli's computations were in no way inimical to the theory of vortices which he proceeded to use to explain the elliptical path of the planets. Johann Bernoulli dropped Descartes' hypothesis that the vortex that carried the planet round the sun was squeezed into an oval shape by the pressure of neighbouring vortices. On Bernoulli's account, a planet, while it was carried by the vortex, had an oscillatory motion to and from the center as a result of its having being originally formed, not at the distance at which it would float in equilibrium in the vortex, but above or below that point. The problem was to demonstrate that the oscillatory motion resulted in an elliptical path, but Bernoulli wisely abstained from claiming more than that the oval might be an ellipse.[14]

It was also necessary to adjust the vortices in such a way that they agreed with Kepler's laws. For the Cartesians this was merely a question of making the speed of each layer of the fluid of the vortex depend in a suitable manner on its radius. The first step had been taken by Huygens who had spoken of the external pressure as manifesting itself as a force varying inversely as the square of the distance from the Center. He was followed by Joseph Privat de Molières who used the inverse-square law to derive Kepler's Third Law, namely that the cube of the mean distances are as the square of the periodic times. This demonstration, Molières asserted, "far from overthrowing the System of Vortices of Descartes, as has been claimed in our days, supports and confirms it".[15]

A year later, in 1729, Molières penned a reply to the Newtonian objection that the particles of a Cartesian vortex could not move in ellipses. He agreed, however, that Descartes' theory had to be modified in one respect: the subtle fluid matter could not be composed of incompressible balls or globes as Descartes had thought. But if it consisted of small elastic vortices, then it could contract in the narrower parts and expand in the wider parts to form an ellipse.[16]

Molières' third essay on this topic is even more interesting for our purposes, since he now proposed to show that Kepler's Second and Third Laws were not only consistent with his theory of vortices, but that they were entailed by it. But whereas we would expect him to attempt to deduce the laws from the properties of fluid motion, he rests his argument squarely on Newton's mathematical demonstrations. Kepler's Second and Third Laws are shown in the *Principia* to follow mathematically from the assumption that the planet is subject to a central force varying inversely as the square of the distance. All Molières felt he had to do was to show that the vortex produced the same effects as Newton's centripetal force. Hence the two theories became indistinguishable in their results! But this did not mean that they were of equal merit. Although the celestial motions could be deduced from Newton's attraction, the virtue of Cartesianism lay precisely in disentangling this beautiful mathematic structure from attraction, and transferring it from a vacuum to a plenum. In this way the doctrine of Descartes made good its claim to be a philosophy of a higher kind. By eliminating attraction, it opposed the revival of "occult properties" or "substantial forms" that had been the bane of scientific progress; by providing some material medium by which a body acted on another at a distance it avoided the reproach urged against the Newtonians that they made a body act where it was not. The clear principles of mechanics could be retained by transferring attraction from a meaningless void to a mechanically intelligible plenum.[17] Even the Newtonians were compelled to admit that Newton himself had not been unwilling to allow that gravity might be an effect produced by some ulterior cause. In this limited sense at least, a Cartesian-inspired quest was not unfaithful to the true spirit of Newtonianism.

The topic for the Essay Prize of 1732 revealed that none of the two rival theories could claim undisputed superiority. The Académie asked to account for the remarkable fact that the revolutions of the planets around the sun, and of the satellites around the planets, as well as their rotation on their axis, including the rotation of the sun, are in the same plane. From the Cartesian standpoint, it was clear that whatever the positions of entry of the planets into the solar vortex, in time the circulating matter would move the planets into the plane of the equator of the vortex. Their problem, therefore, was to explain why the planetary orbits were still inclined at small angles to this plane, and why the comets appeared to be unaffected by the circulation of the vortex. For the Newtonians, however, the problem was considerably more daunting since they had to state why the orbits were concentrated in a narrow zone instead of being randomly distributed. Newton himself could only attribute the rotation of the planets in a common direction in orbits nearly all in the same plane to supernatural agency. Laplace's nebular hypothesis, which may be considered the first Newtonian solution of the problem, was not published until 1796, and, as Whewell remarked, "that hypothesis is, in truth, a hypothesis of vortices respecting the *origin* of the system of the world.[18]

None of the essays submitted in 1732 were judged worthy of the prize, and the same subject was proposed for a double prize in 1734. This time it was awarded to both Johann Bernoulli and his son Daniel. Johann's essay, entitled "Essai d'une nouvelle physique céleste", marks a considerable shift in his thought since his essay of 1730, but he still found the principle of attraction incomprehensible. That is, he admitted attraction as a fact, in the sense that bodies move towards one another, but he rejected attraction as a physical principle or cause. As a disciple of Descartes, he continued to seek clear and distinct reasons for natural events. This did not hamper him from abandoning his earlier explanation of the elliptical orbits of the planets and replacing it by the hypothesis of a perpetual stream of matter tending to the center from all sides. It seemed to Bernoulli that all that Newton derived from the theory of universal gravitation could just as easily be deduced from his own theory which invoked the impulse of a central stream flowing from the circumference to the center of the vortex. The particles of this stream were said to be analogous to those that cause heat and light, and hence their effect (gravity) was inversely proportional to the square of the distance from the center of the vortex. It followed that if the planets had been set in motion in orbits inclined at random, they would describe ellipses about the sun as focus. The irregularity of the directions in which comets move, which Newton had cited against the vortex theory, could now be used as evidence for Bernoulli's revised Cartesian theory. Even comets could be accommodated. Since their orbits are extremely elongated ellipses, for most of their period of revolution they are well beyond Saturn, namely in a part of the vortex where the speed of circulation is so slow that it cannot effectively alter their direction. As a result, the orbits of the comets were not shifted like those of the planets but retained the inclinations they had when they were initially set in motion.[19]

Bernoulli's revamping of the Cartesian doctrine is not unattended by difficulties in its application of the principles of mechanics, but the hypothesis of a central stream could be made to account for the main features of the theory of gravitation. It was even improved in several respects by the eighteenth century Swiss physicist, George-Louis Le Sage. But although the theory could not be refuted in a straightforward way, it gave rise to perplexities of a more philosophical nature. If the world is filled with unending streams of invisible but material and tangible particles, whence do they come, and wither do they go? Or, in Whewell's heightened prose: "Where can be its perpetual and infinitely distant fountain, and where the ocean into which it pours itself when its infinite course is ended? A revolving whirlpool is easily conceived and easily supplied, but the central torrent of Bernoulli, the infinite streams of Le Sage, are an explanation far more inconceivable than the thing explained".[20] I cannot help wondering what Whewell would have said had he known about cosmic rays . . .

The thrust of Daniel Bernoulli's essay was fundamentally Newtonian but even Daniel felt that a vortex had to be retained in the form of a solar atmosphere if the wider cosmological issue, implied by the problem of the plane of the orbits, was to be convincingly addressed. Daniel Bernoulli's solar atmosphere accounted for the present inclinations of the orbits of the planets and comets on the assumption that the orbits had been initially inclined at random, an implication, as we have seen, of Descartes' cosmogony, but in the Newtonian system an arbitrary hypothesis.[21] Attraction remained an obstacle, but the combination of the technical advantages of English physics with the philosophical superiority of French physics had become a goal for which to strive.

Whatever may have been the cause of Daniel Bernoulli's initial hesitation to base his explanation on the mutual attraction and gravitation of bodies, by 1740 he had come to feel that this was the best approach to planetary motion even if the theory of vortices could not be conclusively refuted.[22] It is clear, as E. J. Aiton points out, that:

the choice at this time between the Cartesian and Newtonian systems was not simply a question of deciding between logical alternatives, but involved a subjective judgement after weighing the various arguments for and against the two systems. All the objections to the Cartesian vortices had been known for a long time. On the other hand, as a cause, the attraction was just as unacceptable as it had been for over a century. Yet the success of the Newtonian synthesis was evidently leading many Cartesians to lose faith in the vortex theory and to sense that, despite the philosophical difficulties, the Newtonian system was basically true. They continued to believe that the attraction could be explained by impulsion but reluctantly accepted that there was probably little hope of solving this problem. This may indeed have been the view reached by Newton towards the end of his life.[23]

The Committee of the Académie Royale des Sciences that adjudicated the Prize of 1740 for an essay on the tides acknowledged the trend by dividing the award between four competitors, of which no less than three (among them Daniel Bernoulli) based their explanation on the Newtonian system. The fourth, submitted by Antoine Cavalleri, was Cartesian in spirit but it did not explain the difference between high and low water, as Descartes' had, by the pressure exerted on the ocean by the terrestrial vortex forced into a strait where it passed under the moon. Cavalleri supposed that the waters rose towards the moon and that the terrestrial vortex, being disturbed and broken by the moon, was less effective in forcing them down.[24]

One should not forget that the mention of attraction in the seventeenth or eighteenth century brought to mind magnetism and the
popular illustrations of its mode of operation. Everyone knew that if
iron filings were strewn on a sheet of paper while the two poles of a
magnet were held close below the paper, the filings would arrange
themselves in certain curves, each proceeding from the North to the
South pole of the magnet, like the meridians on a globe of the earth.
We take it for granted (as the good Newtonians we have been brought
up to be) that the direction of these magnetic lines is determined by the
attraction and repulsion of the two poles. But Descartes, and Kepler
before him, thought it was much more natural (and scientific) to infer
that a magnetic vortex constantly streamed out of one pole and into the
other, and hence arranged the iron filings along their lines of flow.
Descartes' account was considered so successful that it was retained
after his planetary theory had entered into a sharp decline. The Triple
Prize awarded in 1744 for the best explanation of the attraction of the
magnet was shared between what J. L. Heilbron has called "embroi-
derers of Descartes' theory".[25] These were no less distinguished figures
than Leonhard Euler, Etienne Francois Dutour, and Johann II and
Daniel Bernoulli.[26] They do not appear to have experienced undue
qualms in being Newtonians in heaven and Cartesians on earth! In the
second edition of his essay entitled "Entretiens sur la cause de
l'inclinasion des orbites des planètes" published in 1748, the French
physicist Pierre Bouguer compares the relationship between thoughtful
Newtonians and Cartesians to political coexistence between rival
political states:

The skilful Newtonian and Cartesian are from now on agreed that attractions must be
looked upon as confirmed by a large number of phenomena that can no longer be
doubted. But these philosophers are not unlike nations that, being unable to arrive at
the bliss of a lasting peace, give themselves, to the best of their ability, the transitory
comforts of a precarious truce.[27]

For Bouguer, attraction was no longer a metaphysical impossibility.
Indeed he was the first to provide experimental evidence of universal
gravitation when, in 1738, he observed the deflection of a plumb-line in
the vicinity of a mountain. Attraction, according to Bouguer, had been
injudiciously ruled out of court because the followers of Leibniz had
made a false application of his principle of sufficient reason. It did
not follow, however, that attraction was a distinct and independent

principle. It could only be contingent upon the principle of impulsion which alone was necessary. But in spite of this aim of supplementing Newtonian mathematics with a physical explanation of gravity, the Cartesians were led to a virtual acceptance of universal gravitation as an axiom. In this they were helped by Bouguer and others who no longer saw in attraction a threat to the mechanical philosophy.

The abandonment of planetary vortices did not entail, as I have mentioned, the immediate demise of the model of a fluid vortex for the explanation of physical phenomena. Leonhard Euler remained convinced that gravity had a physical cause, not yet known in detail but arising from the action of the fluid matter filling space.[28] Moreover Euler was the first to provide a sound theoretical basis for the construction of fluid models of the aether that Helmholz was to develop in the nineteenth century.

V. CHANGE BY TRANSFORMATION

The protracted debate between the Cartesians and the Newtonians illustrates Whewell's contention that when different and rival explanations of the same phenomena are held, the defeated hypothesis is not always suppressed in a theoretical paradigm-shift. It is often penetrated, infiltrated and transformed until it gradually passes into its technically superior rival. Each case has to be taken on its own merits, but scientific change need not be made synonymous of upheaval. In a genuine sense, an early theory can contain the seeds of a later one. The vortex theory embodied two mechanical principles that Newton was to use in arriving at his own planetary theory, namely the principle of inertia and the concept of centrifugal force. The philosophical background against which the theories were constructed played a role that cannot be neglected. The epistemological stance of Cartesians and Newtonians were profoundly at variance. Whereas the former always looked for hidden mechanisms, the latter shied away from speculation about the ultimate cause of phenomena. Yet the Newtonians did not find action at a distance objectionable although it was unintelligible in terms of everyday experience of bodies in motion. The Cartesians did not deny attraction as an empirical occurrence but they argued that the task of the natural philosopher was to delve below the level of mere appearances. Newton pondered the nature of gravity only to reach the conclusion that it was sufficient if it could be deduced (in some unspecified way) from the phenomena. Universal attraction struck the Cartesians as an occult property and they instinctively rejected it. When the practical advantages of the Newtonian system became apparent, the Cartesians used it, initially as a mere mode of computation, and later as a provisional explanation to be eventually superseded by a fuller account in terms of impulsions.

When we judge the manner in which the Cartesians tackled the motion of the vortices we should compare their results not with Newton's dynamics of particles but with his own treatment of vortices, which is hardly an improvement on the best work of the Cartesians. Newton, like the followers of Descartes, considered the fluid strata as though they were solid rims, and turned them into the spheres of ancient astronomy. When the Cartesians realized that the dense fluid postulated by Descartes was indefensible, they tried to reconcile the mathematical results of Newton with their own theory by postulating a fluid of negligible inertia. They strove mightily to explain the effect of gravity. A new kind of impulsion was needed, and when it was not forthcoming, the quest for something behind attraction declined. The accurate mathematical representation that the Newtonians gave of the planetary motions came to be accepted by most as the best that could be hoped for. By most but not by all. For despite the positivist advocates of the elimination of causality in physics, the notion of a spatial continuity of causes continued to have its champions. Attempts to explain gravity without action at a distance culminated in Einstein's general theory of relativity employing a model very different from the one of the Cartesians yet capable of putting their minds at rest. As E. J. Aiton puts it: "Gravity, conceived as a force, an attraction acting at a distance, was after all an occult quality that could be eliminated by geometrizing".[29]

Whewell's view of the manner in which rival theories pass into one another is not fashionable because contemporary thinking on scientific change has swung away from the naive block-building metaphor of the positivists to an equally naive analogy borrowed from a superficial account of political revolutions. Upon closer inspection radical political breaks are seen to have both assignable causes and a striking continuity with the earlier state of affairs. The ascription or leaps of faith to scientists is a consequence of a tendency to think, after the event, in dichotomous categories. We look at the prolonged controversy between Cartesians and Newtonians and we say: "X either held that planets moved in vortices or that they obey the same laws that govern falling bodies on earth. He could not have maintained both positions without contradicting himself". But as we have seen Cartesians held both views and with good reasons. Or again, it might be said, with the same apodictic tone of voice, "Either the planets move in their orbits according to the laws that Johann Bernoulli based on the Cartesian theory, or they obey the laws laid down by Newton". The logic of the situation, however, is really as follows: "If the hypothesis of Bernoulli is correct, it is so because it agrees in its result with the Newtonian theory. It is not only possible that both opinions may be true, but it is certain that if the first one is, so is the second".

Alternatives of this kind have always been the mainstay of philosophers and they were once invoked to enable us to see on what side rationality lies. When it became clear that the evidence spoke with a forked-tongue, it was no longer possible to claim rationality for one side only. Hence the appeal to the revolutionary leap to account for the apparenly instantaneous (but really gradual) transition from one alternative to the other. The progressive Newtonization of continental Cartesians has nothing irrational about it. They looked, and they tell us in considerable detail why they looked, before they leapt.

McGill University

NOTES

* The author wishes to thank the Social Sciences and Research Council of Canada and McGill University for generous support of his research.
[1] Max Born. *The Restless Universe*. Translated by W. M. Deans. New York: Dover, 1951, p. 73.
[2] Karl R. Popper. *The Logic of Scientific Discovery*. London: Hutchison, 1959, p. 111.
[3] "For my part I do, qua lay physicist, believe in physical objects and not in Homer's gods; and I consider it a scientific error to believe otherwise. But in point of epistemological footing the physical objects and the gods differ only in degree and not in kind. Both sorts of entities enter our conception only as cultural posits". Willard Van Orman Quine, "Two Dogmas of Empiricism" in *From a Logical Point of View*, 2nd ed. New York: Harper & Row, 1961, p. 44.
[4] Galileo Galilei. *Dialogue Concerning the Two Chief World Systems — Ptolemaic and Copernican*. Translated by Stillman Drake. Berkeley and Los Angeles: University of California Press, 1962, p. 328 (in the national edition of Galileo's *Opere*, edited by A. Favaro, Florence: G. Barbèra, 1899—1909, Vol. VIII, p. 355.

⁵ *Transactions of the Cambridge Philosophical Society* **9** (9 May, 1851), pp. 139—147. Reprinted in Robert E. Butts (ed.) *William Whewell's Theory of Scientific Method.* Pittsburgh: University of Pittsburgh Press, 1968, pp. 251—262, from which I quote.
⁶ *Ibid.*, p. 251
⁷ René Descartes. *Principia Philosophiae*, Part III, art. 52—147. In C. Adam and P. Tannery (eds.) *Oeuvres de Descartes*, 12 vols and index, 1897—1913. Reprint, Paris: Vrin, 1956—1973, vol. VIII-l, pp. 105—202.
⁸ *Ibid.*, Part IV, art. 24—27, pp. 214—216. An excellent account of the vortex controversy is to be found in E. J. Aiton. *The Vortex Theory of Planetary Motions.* London: Macdonald, 1972.
⁹ René Huygens. *Oeuvres Complètes*, 22 vols. La Haye: Martinus Nijhoff, 1888—1980, vol. XXI, p. 79. See Aiton, *op. cit.*, pp. 75—78.
¹⁰ Whewell, *art. cit.*, p. 254.
¹¹ Huygens, *op. cit.*, vol. XXI, pp. 472—473.
¹² Joseph Saurin. "Démonstration d'une proposition avancée dans un des mémoires de 1709 . . .". *Mémoires de l'Académie Royale des Sciences.* Paris, 1718, pp. 191—199. See Aiton, *op. cit.*, pp. 172—176.
¹³ The original memoir of 1669 is printed in the *Oeuvres*, vol. XIX, pp. 631—640. It achieved notoriety, however, in 1690 when it was included, with additions, as a supplement to the *Traité de la Lumière, Oeuvres*, vol. XXI, pp. 451—488.
¹⁴ Johann Bernoulli. "Nouvelles pensées sur le systéme de M. Descartes". In *Opera Omnia*, Geneva, 1742, reprint Hildesheim: Georg Olms, 1968, vol. III, pp. 132—173.
¹⁵ J. P. de Molières. "Loix générales du mouvement dans le tourbillon sphérique". *Mémoires de l'Académie Royale des Sciences.* Paris, 1728, p. 265. See Aiton, *op. cit.*, pp. 209—214.
¹⁶ J. P. de Molières. "Problème physico-mathématique, dont la solution tend à servir de réponse à une des objections de M. Newton contre la possibilité des tourbillons". *Mémoires de l'Académie Royale des Sciences.* Paris, 1729, pp. 235—244.
¹⁷ J. P. de Molières. "Les loix astronomiques des vitessess des planètes dans leurs Orbes, expliquées méchaniquement dans le Système du Plein", *Mémoires de l'Académie Royale des Sciences.* Paris, 1733, pp. 301—312.
¹⁸ Whewell, *art. cit.*, p. 348.
¹⁹ Johann Bernoulli. "Essai d'une nouvelle physique céleste". In *Opera Omnia*, vol. III, pp. 261—364.
²⁰ Whewell, *art. cit.*, p. 258.
²¹ Daniel Bernoulli. "Recherches physiques et astronomiques". *Recueil des pièces qui ont emporté le prix de l'Académie Royale des Sciences*, Paris, 1752, vol. III, pp. 93—122. See Aiton, *op. cit.*, pp. 235—239.
²² Daniel Bernoulli. "Traité sur le flux et reflux de la mer". *Ibid.*, vol. IV, pp. 56—57.
²³ E. J. Aiton, *op. cit.*, p. 247.
²⁴ Antoine Cavalleri. "Dissertation sur la cause physique du flux et du reflux de la mer". *Recueil des pièces qui ont emporté le prix de l'Académie Royale des Sciences.* Paris, 1752, vol. IV, pp. 21—28.
²⁵ J. L. Heilbron. *Elements of Early Physics.* Berkeley: University of California Press, 1982, p. 23, n. 4.

[26] *Recueil des pièces qui ont emporté le prix de l'Academi Royale des Sciences.* Paris, 1752, vol. V. pp. 1—144. The three essays are separately paginated at the end of the volume.

[27] Pierre Bouguer. "Entretiens sur la cause de l'inclinaison des orbites". *Recueil des pièces qui ont emporté le prix de l'Académie Royale des Sciences,* Paris, 1752, vol. I, p. 48.

[28] Leonhard Euler. *Opera Omnia.* Series tertia, volume I, Leipzig and Berlin, 1926, p. 10.

[29] E. J. Aiton, *op.cit.,* p. 262.

MARCELLO PERA

THE RATIONALITY OF DISCOVERY:
GALVANI'S ANIMAL ELECTRICITY

1. THE SOCIETY OF DISCOVERY AND ITS ENEMIES

The "friends of discovery" have only had the briefest time in which to reconstitute themselves and already a new challenge has reached them. The old challenge said that what they are looking for, a logic of discovery, does not exist. The new challenge is stronger. It says that, even if they reached their aim and proved that scientists often reason from data *to* theories or hypotheses, this would be irrelevant because the eventual generative arguments wouldn't have epistemic strength *per se*.

To start with, let us define "discovery". Discovery is a process of thought that leads from (at least partially) observational premises to cognitive conclusions, generally in the form of laws or theories. Another appropriate term is "invention". According to the friends of discovery, the process of invention involves an inference that, linguistically, takes the form of an argument.

This clarification is not enough, however. Larry Laudan attacked those who defended the idea of a rule-governed process of discovery first by asking for more detailed explanations. He wrote that

the older program for a logic of discovery at least had a clear philosophic rationale: it was addressed to the unquestionably important philosophical problem of providing an epistemic warrant for accepting scientific theories. The new program for the logic of discovery, by contrast, has yet to make clear what philosophical problems about science it is addressing.[1]

Laudan is right. The aim of the friends of discovery is not always clear: do they only intend to reconstruct logically the context of invention (which would be an answer to the old challenge) or, instead, do they intend to show that this context has its own epistemic value (which would be the appropriate answer to the new challenge)? Of course, the society of discovery should reach both aims, especially the second one. Only if the context of invention has epistemic value, does it make sense to work for its reconstruction, analysis and justification. In the opposite

177

J. C. Pitt and M. Pera (eds.), Rational Changes in Science, 177–201.
© *1987 by D. Reidel Publishing Company.*

case, it can be left to whoever took care of it up to now: usually psychologists, sociologists or historians.

Even with this clarification, Laudan is not satisfied. He is skeptical about the possibility of reaching the second aim. His view is that "in whatever guise the logic of discovery may eventually appear, it seems clear that its traditional *epistemic* role has been well and truly pre-empted by the theory of testing"[2]. Laudan does not give many reasons for his skepticism; he believes that it is the responsibility of the friends of discovery to prove that a logic of invention has epistemic value *per se*, or to prove that it can contribute to any one of the numerous problems that cannot be solved by the logic of justification (for instance, problems regarding the heuristic).

Thomas Nickles argues that the friends of discovery have not met Laudan's challenge. He looks forward to a post-Laudanian era when the central matter will be "the logic of discoverability" irrespective of the truth of Laudan's view. This is intended not as the logic of the original or effective invention of a cognitive claim, but as a logic of how that claim "*could have been* discovered in the rationally specified manner had the necessary information and analytical techniques been available."[3] According to Nickles, there are no logical reasons to believe that the effective process of discovery must be an inference, and, even if it were, "there need be no particular relation between justification and the actual, original mode of generation. Discoverability need not overlap discovery either temporally or logically".[4]

I believe that the notion of discoverability constitutes an important addition to the discussion. As logicians, we are interested in articulated arguments. The original reasoning behind a discovery can be inaccessible even to the investigator himself. And since it may also be the case that an investigator arrives at a hypothesis H by means of a different route than that used by the individual who originally generated it, the exhibition of an argument about how we *could* discover H is equally satisfactory for the friends of discovery. However, this does not mean that some original argumentative process does not exist, or that it is irrelevant, as Nickles maintains.

As I argued elsewhere,[5] generative arguments are indispensable. They give hypotheses that degree of initial plausability without which, according to the Bayesian schema, they cannot receive even a degree of "final" confirmation. Nickles objects that this line of defense proves at most the existence of *de facto* connections, necessary to explain the *efficiency* of scientific research, but not *de jure* connections, necessary to prove its *possibility*. But this objection is not convincing for both a factual and a logical reason:

(a) By separating the original discovery from discoverability, Nickles presumably reconstructs research in the following kind of sequence: (1) ascertainment of certain data D_0; (2) free intuition of an hypothesis H able to explain D_0 (3); discoverability arguments from D_0 to H; (4) eventual consequentialist arguments from H to $D_1 \ldots D_n$. The crucial question here concerns: which hypothesis does the scientist take into consideration? Or does he consider all of them, even the most crazy? Nickles says that scientists "if they wish, can evaluate as contributions to their current research problems the marks produced by my six-year-old daughter or by the cat running across the computer keyboard".[6] However, it is a *de facto* truth that scientists *do not* want to do so.[7]

(b) Even if scientists wanted to they couldn't do it. Science is not simply research; it is *efficient* research. If we leave out efficiency, we miss the difference between, say, how a magician and an astronomer predict the future. Success would turn out to be unexplainable and miraculous, because to discover something (or to solve a problem) by simply developing or testing every scrawl or every possible psychological *Einfalle* is a miracle considering the range of possibilities. On the contrary, if we do not break "any logically necessary connection between discovery and justification",[8] efficiency may be explained as deriving from the fact that scientists take into consideration, develop, and test only those solutions of problems for which logical connections may be shown to exist with the available data. Then, these connections, at least the first ones, can be considered to constitute the argument, or part of the original argument, behind the discovery.

With this clarification we can now come back to the most important matter proposed by the new challenge to the society of discovery: do the arguments of discovery (the original or their discoverability substitutes) have their own epistemic value? Or are they completely absorbed by those of the consequential test as Laudan suggests? This essay aims at maintaining that the first have epistemic value as do the second, and that a logic of discovery can help solve some of the problems that a logic of justification *post hoc* does not even allow to arise. This logic will combine the advantages of the old *Novum Organon*, the traditional logic of discovery, and the advantages of *Novum Organon Renovatum*, the traditional logic of justification, without carrying too many of their disadvantages. We can define this combination as *Novum Organon Reformatum*. The first thing is to analyze why this new reform of the *Organon* is necessary and of what it consists.

2. THE POVERTY OF GENERATIVISM AND CONSEQUENTIALISM

For our purpose, we can represent the defenders of the *Novum Organon*, or generativists, as committed to the view that the process of scientific inquiry has two fundamental characteristics: (i) it is argumentative, that is, expressible in arguments with premises and conclusions supported to some degree; (ii) it is unitary, that is, not logically distinguishable into "contexts". It follows, therefore, that the process is governed by only one kind of rules: those of the logic of *discovery*.

From Bacon to Hume and Mill and even after them, the aim of the generativist program has been to analyze and justify these rules. Of course, the program has its variants. For example, one version predicts that the initial data from which discovery starts are merely observational; this is the case of Bacon's "induction from phenomena" or Mill's "canons of discovery". In other versions the program predicts even stronger rules of inference, as in Newton's "deduction from phenomena", or it accepts theoretical elements among the initial data, or it allows the degree of implication of the inferred conclusion to be lower than one. None of these variations need be examined here; the program can be evaluated on the basis of only its fundamental characteristics.

Certainly generativism offers some advantages. For example, it explains better than consequentialism the efficiency of research, our learning from mistakes and from new results, or the apparent almost inductivist, ever-generalizing or unifying, growth of science. But these advantages have a high cost. The following are the most important disadvantages.

(1) Generativism includes the rules of starting from known data, but it does not specify *which* data. It is thus easy for hypothetico-deductivists to reply that it is not possible to observe or gather data and then generalize or infer conclusions, unless there are theoretical selection criteria for the premises. As P. Medawar said "we cannot browse over the field of nature like cows at pasture".[9]

(2) Generativism finds itself in trouble trying to explain the fallibility of research. If there are rules that lead from observational data to

theoretical conclusions, how can we arrive at false conclusions starting from true data and applying the rules correctly? This objection also hits the weak variants of the program. If D_0 represents the initial data, R an inductive rule applied to D_0, and H an inductive conclusion, then the degree of support or partial truth of H should not decrease or fall to zero with the acquisition of new data $D_1 \ldots D_n$.

(3) Generativism leaves too little room for individual sagacity. Bacon thought that the rules of his *Novum Organon* worked so sharply as "to level man's wits". Mill did not have a lower opinion of his own rules. However, it is hard to understand how investigators can arrive at divergent conclusions by correctly applying the same rules of discovery to the same data. Hume's answer, that the rules of discovery "are very easy in their invention, but extremely difficult in their application",[10] solves the problem at an exorbitant price for a generativist, because it leaves him without any instructions for generating cognitive claims.

(4) Finally there is another important disadvantage. For generativists, invention and justification coincide. A conclusion H, generated by data D by applying the rules of discovery, is also a justified conclusion; therefore new data $D_1 \ldots D_n$ add little or nothing to the truth value of the established conclusion. From the generativist's point of view, there is no reason to believe that a new datum predicted by H confers on H more credit than that conferred by the data already known at the moment of the construction of H.[11] However, our intuition is that more positive data confer greater value on H.

About the middle of the XIX Century, the generativist program — for reasons which need not be examined here — was superseded by the consequentialist program of the *Novum Organon Renovatum*. For our purposes this program can be taken as committed to the view that the process of research is (i) partially argumentative; (ii) logically non-unitary. The second characteristic is intended in the sense that the whole process is divided into at least two completely different episodes: invention, where the cognitive claim is first advanced, and justification, where the claim is tested. It is assumed that the first episode is a-logical and, even if logically reconstructible, epistemically irrelevant *per se*. It is also assumed that the arguments of the second episode are of consequential type, that is, they proceed *from* the claim to its logical consequences. Therefore, the consequentialist program, contains only one kind of rules: those of the *logic of confirmation*.

Like the generativist, the consequentialist program has its variants.

One version predicts the inductive retro-conversion of verified consequences, in the sense that these assure, under special conditions, the
truth of the hypothesis. There are also weaker forms of inductivism, as
in the case of modern versions such as Carnap's. Or there is the simple
deductivist account, which has an even weaker claim to truth, as in
Popper's version. However, in all these variants the program keeps the
two characteristics outlined above, and, from them, it derives its
advantages. First, consequentialism can explain much better than
generativism the fallibility of research; no hypothesis confirmed by the
hypothetico-deductivist method can ever be considered true. Moreover,
consequentialism makes room for the tentative aspect of scientific
research; hypotheses are not mechanically generated by rules, rather,
they are suppositions that require skill and imagination. Lastly, consequentialism explains better than generativism the theoretical disrup-
tions of tradition; it doesn't exclude continuity, but it presents the big
revolutions in a very natural way, as a result of audacious attempts to
change the ideas received and to observe the same facts in the light of
different hypotheses or different points of view. There are defects too.
The following points (1*)—(4*) correspond to the previous problems
(1)—(4) with generativism.

(1*) Consequentialism prescribes that one must guess at hypotheses
to solve cognitive problems, but it does not give any instructions about
which hypotheses. As we have already mentioned, scientists, for reasons
of intellectual economy, cannot consider all hypotheses equally credible.
But, if all hypotheses are suppositions, or "wild" guesses, do any have
greater initial credibility? In a similar view consequentialists do not
have much to say about heuristic (their only meagre advice being
"invent and test."). They could introduce selective filters or bonds in
order to limit the "free creation" of hypotheses; but in this way they
would admit that, contrary to their own assumption, the success of
research depends on factors which belong to the context of discovery.
Moreover, a bound creation is similar, if not identical, to a creation
generated by rules or achieved by following rules.[12]

(2*) Consequentialism meets difficulties explaining the frequent
success of research. Why do our "attempts to guess" have such a
success? Whewell or Popper's view that hypotheses are "happy guesses"
or "lucky guesses" is not an explanation. What is required are the
reasons for so much "happiness" and "luck". The naturalistic thesis of
the adaptation of mind to the environment works better but it is also

generic. Moreover, a mind more prone to produce answers suitable to the environment looks like an instructed mind; in this way success would be tied to the techniques of instruction and hypotheses generation. It should also be noted that a coherent consequentialist like Popper even prohibits raising the problem of success.

(3*) While it insists on binomial "suppositions and confirmations" or "conjectures and refutations", consequentialism does not explain our learning from experience. Its favorite principle is *that* we learn from our mistakes; but it is not able to say *how* our mistakes lead to adjustment in hypotheses. In Popper's evolutionary epistemology the generation of every new idea is independent of the mistakes discovered in the previous ideas, it is "blind" like a biological mutation. Nevertheless, current scientific practice shows that new ideas often derive from the available data and that the latter instruct the former.

(4*) Finally, consequentialism reduces, if not nullifies, the probatory value of the *known* data. It requires "novel facts" be derived from an hypothesis (in one of many different senses, temporal, heuristic, etc. of "novel"); in the opposite case, the hypothesis is suspected to be *ad hoc* and credited with weak empirical support. This has unpleasant consequences. In order to establish if an hypothesis is *ad hoc*, one would have to search the private biography of the investigator or do a complete inventory of his background knowledge. The degree of confirmation of an hypothesis would depend on psychological, chronological, or cultural circumstances. But a pragmatic theory of confirmation is not available; and even if it were, it would transform, in an unacceptable manner, an essential assumption of the consequentialist program: judge an hypothesis only on the basis of its logical consequences.

Let us summarize and compare. From (1)–(4) and (1*)–(4*) it follows that the generativist program and the consequentialist program have two symmetrical kinds of defects. Descriptively, considered as reconstructions, they are not completely adequate to scientific practice; both of them neglect relevant characteristics of the *explicandum*. Normatively, considered as systems of techniques or rules, they make it difficult to reach the aims they pursue, or to understand why these aims are not always reached.

In matters such as these, we cannot hope to obtain all the advantages and no disadvantages. However, we can try for the best possible results. In this case, a result would be better if it allowed us to justify a method satisfying at least two characteristics: it must be fallible and it must give

some indications about the problem of the heuristic. This essay intends
to contribute to this result. In particular, it aims to defend the four
following theses.

(A)   There are inferences from known (empirical or theoretical) data
      to cognitive claims (hypotheses). The generation of an hypothesis
      is not a spontaneous and miraculous process.
(B)   Logically, the inference of discovery is any inductive or deductive
      or argumentative (dialectical) inference from data to an hypothesis.
      That means there is no *special logic* of discovery and no *logical*
      distinction between contexts. There are only pragmatic and func-
      tional distinctions which refer to the direction or aim of the
      inference. This allows us to introduce not only two but *more* than
      two epistemic contexts, all distinguishable in the same pragmatic
      and functional way.
(C)   The logic of discovery has an epistemic value *per se*. This means
      that the reasoning which leads to the introduction of an hypothesis
      gives it a justification which is independent from, and not
      absorbed by, the *post hoc* test of the hypothesis.
(D)   The inference from data to hypotheses is guided by heuristic
      principles that are both situational (in that they are tied to the
      specific methodological principles of the program or of the
      research tradition in which the inference takes place), and general
      (that is, dependent on permanent or long term models of argu-
      mentation).

If these four theses can be proved, they present an image of a
method that combines aspects of both the generativist and consequentialist programs. As the *Novum Organon*, this method treats the
generation of an hypothesis as a rational matter (theses A and B); as the
Novum Organon Renovatum, it does not reduce the epistemic foundations of hypotheses to only the generative act (thesis C). Moreover it
intends to show that scientific rationality is not a-historical because
research is also guided by situational principles (thesis D).

In order to discuss these four theses, I first will develop an historical
example and then derive some morals from it. As a *case study*, I will
use the discovery of "animal electricity" by L. Galvani.[14]

3. GALVANI'S DISCOVERY OF ANIMAL ELECTRICITY

On January 26, 1781 Luigi Galvani had a "casual" experience which proved to be important in the history of physiology and physics. The situation in question, known as "Galvani's first experiment", was very simple. He noted that if, with an anatomical knife, he touched a frog "prepared in the usual way" (that is, cut at the height of the spinal cord, without skin and with the crucial nerves uncovered, see Fig. 1), it contracted at the same time a spark was obtained from an electrostatic machine located a certain distance away.[15]

Fig. 1. Galvani's "frog prepared in the usual way" (autograph from *Memorie ed esperimenti inediti di L. Galvani*, Cappelli, Bologna 1937).

At that time, the idea that there was electricity in animal bodies was rather popular but without any serious evidence. Galvani was sure about it and was looking for solid proof. Although the phenomenon he observed that day was not of this kind, it was not irrelevant because the contractions, obtained without any direct discharge on the nerves of the frog, objectively increased the suspicion about a connection between animal physiology and electricity.

After more than five years of research, Galvani had another "casual" important experience. At the beginning of September 1786 he took one of his frogs out onto a terrace to see whether it contracted when lightening flashed. After putting a hook (which happened to be made of brass) in the spinal cord of the frog and hanging it on the iron railing, he observed that when its legs touched the railing the frog contracted. This was "Galvani's second experiment" (Fig. 2).

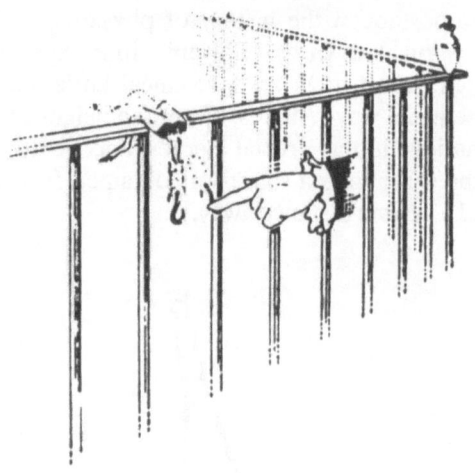

Fig. 2. Galvani's "Second Experiment" (from M. Sirol, *Galvani et le galvanisme*, Vigot Frères, Paris 1939).

It certainly proved much more than the first one: it made probable the hypothesis that there was an electric fluid in the frog, or a fluid very similar to the electrical one, because the frog contracted in the same way it did when hit *directly* by an electric discharge. Galvani's argument can be reconstructed on the basis of what he wrote in a memoir published a while after his experiment.[16] This argument results from the following *structural analogy*:

I.   In situation S_1 (circuit: Brass Hook-Frog-Iron Railing) muscular contractions are produced.
      The contractions observed in S_1 are the same as produced when electricity is discharged directly on the nerves of the frog.
      The contractions produced by electric discharges on the nerves are due to the passage of an electric fluid.

      H: It is probable that the contractions of the frog are due to a passage of electric fluid.

The plausibility of H was reinforced by the fact that the contractions took place also in a laboratory with circuits made of different kinds of metals (situation S_2), which excluded their dependence on an unperceived discharge of atmospherical electricity. Nevertheless, couldn't there be "another principle" beside the electric one?

In order to exclude this possibility, Galvani invented an experiment. He put the frog's nerve and muscle in contact with non-conductor materials (situation S_3). Obviously his argument was: if the contractions take place in S_3 as in S_2, then H is false, but if they do not take place, then there is one more reason to believe in H. In other words, Galvani made the following *eliminative induction*:

II.   In S_2 contractions take place.
      The difference between S_2 and S_3 is the passage of electric fluid, only possible in S_2.

      H: It is probable that the contractions of the frog are due to a passage of electrical fluid.

However, there was another question that couldn't be answered by I and II. Was the electrical fluid internal to the frog, or was it external, due to some unknown but conceivable actions of metals or of something else? Although Galvani had no direct proof to solve this problem, he made a valid indirect argument; namely, if what physicists maintained was true, in particular, that no opposite charges can exist on metals and that metals function merely as conductors for the fluid, then opposite electric charges had to exist in the frog, and, therefore, the fluid that contracted the muscles had to be intrinsic to the frog. So, Galvani made another *eliminative induction*:

III. In S_1 contractions take place.
      The difference between S_1 and S_2 is the passage of the electric fluid only possible in S_2.
      The metallic arc functions merely as a conductor for the fluid.

      H: It is probable that an intrinsic electric fluid exists in the frog.

The plausibility of the conclusion of this induction was also increased by a positive argument. While metals with double and opposite electricity do not exist, there is an instrument in which it does exist, namely the Leiden jar. The frog and the Leiden jar have an important property in common: both discharge with the application of a metallic arc. Galvani thus reinforced III with a *structural analogy*:[17]

IV. The frog contracts when a metallic arc connects its nerves and muscles:
The Leiden jar discharges when a metallic arc connects the external armature and the internal one.

H: It is probable that the frog is an organic Leiden jar.

But other analogies proved to be possible as well; for example, with the well known "electric fish" (like the torpedo) or with other instruments, like the electrophor, or with minerals like tourmaline.[18] Therefore, Galvani's total argument had the structure of the following *structural analogy*:

V. The physiology of the frog has the properties $P_1 \ldots P_n$. The electric fish (or the electrophor or tourmaline) have the properties $P_1 \ldots P_n$.

H: It is probable that double and opposite electricity exists in the frog.

Galvani's subsequent research was addressed to enlarging the analogical basis of his hypothesis trying to find other properties common to the frog, electric animals, electric instruments. Of course, the Leiden jar was preferred in these new experiments because it was the condenser with the best known structure and laws. One day, while he was carrying on different experiments with the jar, he noticed that the electric "flake" which appears on the charged bottle, animal respiration, and the flame, had some properties in common.[19]
Let us consider the properties of the flake. Galvani observed at least nine of them. They are: P_1: under a glass bell, the flake weakens and eventually disappears. P_2: This weakening is proportional to the dimension of the container. P_3: If we make a hole in the container, the flake regains strength. P_4: After it is extinguished, it can be spontaneously revived by taking the bottle outside. P_5: It is extinguished if the air is mephitic. P_6: If it is extinguished because of the impurity of the air, it does not revive. P_7: In a vacuum its rays decrease in number and increase in length, become irregular and finally disappear. P_8: It pollutes the air. P_9: If it is weak, it is extinguished with few blowings; if it is strong, with vigorous blowings.
Since these properties seemed to correspond exactly to properties of the flame and animal respiration, Galvani concluded that a resemblance

should exist "between the phenomena and principles from which those facts derive".[20] Therefore, his reasoning was another *structural analogy*, this time stronger than V, because of the increase in the number of properties taken in consideration:

VI. Electric flake, flame and respiration share properties $P_1 \ldots P_n$.
     The "active principle" of the flake is the electrical fluid.

H: It is probable that flame and respiration depend on the same "active principle".

Clearly this analogy did not have a direct influence on the problem of animal electricity. However, it did have something indirect to say about it. If it is reasonable to suspect the existence of a cause or active principle common to respiration, flake and flame (because they share the same properties), and if it is reasonable to suspect that this is an electric principle (because the principle of the flake is) then it is also reasonable to suspect that an electric fluid develops inside the human body.

Another property of the flake became even more important later. Continuing to work with the Leiden jar, Galvani discovered that the flake revived when a spark was produced by the first conductor of an electric machine situated at a certain distance, exactly as it had happened in his first experiment with the frog. The phenomenon was as new and unexpected as the other was at its time. It was reasonable to suppose that both were governed by the same principle, considering the many properties they had in common. The reasoning was simple: if the discharge produced by an electric machine revives the flake of the jar, then it extracts the fluid from it; in the same way, if the discharge revives the frog, by making it contract, it must extract or excite its own fluid. Again, Galvani made a *structural analogy*:

VII. The Leiden jar presents the new property P_i.
     The muscular fiber presents the property P_i.

H: It is probable that the muscular fiber is an organic Leiden jar.

At this point, we can leave Galvani's research. The *Commentary* he published in 1791, *De Viribus Electricitatis in Motu Musculari Commentarius*, did not contain new evidence for the hypothesis of animal electricity, but the great majority of the scientific community considered the arguments listed above powerful enough to transform a conjecture with little or no real support into an acceptable theory. For example, Alessandro Volta (who later on rejected Galvani's theory giving rise to a famous controversy with him) after reading the *Commentary* declared himself a "fanatic"[21] follower. However, before we leave the research, let us consider some points which will be useful later. We know some dates of the experiments, but not all of them. The inferences I—V are in strict chronological order and they all belong to September 1786. We are not sure about inference VI, but almost certainly it too belongs to 1786. Inference VII is not certain either; in the *Commentary* Galvani describes it as "recent",[22] but this is not enough to locate it with precision. However, we can say that VII is related to VI: they belong to the same kind of experiments, and surely VII follows VI because, when Galvani refers to VI, he does not mention VII. Since reliable documents do not exist, different chronological sequences are therefore possible. The following ones are the most probable:

Sequence A: VI, VII, I—V. There would be two cycles of experiments (VI—VII and I—V) located in 1786; because of its similarity with the "first experiment", VII would function as a stimulus not weaker than I.

Sequence B: VI, I, . . . VII . . . V. In this case the stimulus would belong to VI and I; VII would be located between 1786 and 1791, and it would reinforce the belief developed along with the preceding inferences.

Sequence C: I—VII. The cycle I—V would raise the first suspicion and reinforce it; VI—VII would then provide proof by giving indirect but independent support. In this case VII would also be located between 1786 and 1791.

If we interpret literally Galvani's remarks in the *Commentary*, sequence C is, historically, the most probable. But, neither of the other two sequences can be rejected. Later we will see that this circumstance has some importance for the logic of discovery.

4. A LOGICAL ANATOMY OF GALVANI'S DISCOVERY

One swallow does not mean Spring has arrived. We cannot derive overly general and ambitious conclusions from only one case. However, our case can be used to *illustrate* theses (A)—(D) of the *Novum Organon Reformatum.*

Let us consider thesis A. Galvani's research clearly proves that there is an inference from data *to* hypotheses. But not merely from empirical data. In order to arrive at his hypothesis, Galvani needs additional premises, as is shown by a detailed examination of some of the steps of his strategy. For instance, with the second experiment Galvani confronts two facts: the fact O_1 that when a circuit of a certain kind is formed, contractions in the frog are produced; and the fact O_2, that these contractions are identical to those obtained with artificial and atmospheric electricity. Now O_1 and O_2, in addition to a known law (L_1: electricity contracts animal tissues), make H probable. However, these premises are still not enough. In addition to O_1, O_2 and L_1, Galvani needs at least another law (L_2: metals are good conductors of the electric fluid); a physiological theory (T_1: muscular contractions are due to the mechanic action of a fluid); and a methodological assumption (A_1: equal effects derive from the same cause). So, inference I presents the following structure:

$$O_1, O_2$$
$$L_1, L_2$$
$$T_1$$
$$A_1$$
$$\overline{}$$
$$H$$

The same happens for inference II. Here, in order to infer H, Galvani needs certain observational facts, namely O_3: in the experimental situation S_1 contractions take place; and O_4: contractions do not take place in situation S_2. Moreover, in addition to L_2, T_1 and A_1 already applied, he needs L_3: non-conductor materials do not allow the electric fluid to pass, and T_2: electricity is a thin fluid. So, inference II presents the following structure:

$$O_3, O_4$$
$$L_2, L_3$$
$$T_1, T_2$$
$$A_1$$
$$\overline{}$$
$$H$$

But the complete structure of Galvani's strategy is more complex still; it can be expressed in the following way:

$$O_1 \ldots O_n$$
$$L_1 \ldots L_n$$
$$T_1 \ldots T_n$$
$$A_1 \ldots A_n$$
$$I$$
$$\overline{}$$
$$H$$

This inference means: if certain observational facts $O_1 \ldots O_n$ are verified, and certain laws of physics $L_1 \ldots L_n$ are well confirmed, and certain theories of physics and biology $T_1 \ldots T_n$ are reliable, and certain general assumptions regarding nature $A_1 \ldots A_n$ are maintained, and a specific assumption I about the phenomena of the frog is made, then the hypothesis of animal electricity follows as an (inductive) consequence. Since premises L and T were taken for granted (they belonged to the truths recognized by the scientific community), Galvani's efforts were concentrated in trying to enlarge his observational basis in order to show how incremental increases in the probability of H would correspond, *legibus paribus*, to each increase in facts. For the success of these attempts, assumption I was essential. This is a particular interpretative theory according to which the phenomena of the frog belong to the class of organic phenomena and, therefore, must be explained in organic terms. As we will see, I plays an important two-fold role, both heuristic and validating.

Summing up, Galvani's strategy proves that he does not treat his hypothesis in the way advised by consequentialists; rather, he reasons from facts *to* hypothesis. In most cases, as inferences I–VII show, he makes use of *inductive* arguments: the As, are typical principles of induction that allow him to extend a certain conclusion to unknown cases. An example is the assumption A_1 already met; another is the assumption of the proportionality of causes to effects which lies behind the eliminative induction III. Here, in order to exclude the hypothesis that the metallic arc is active and causes the contractions of the frog, Galvani argues that, since the arc is "very thin and very short" it cannot "excite those rather strong contractions";[23] an argument that obviously works only if one is willing to admit that effects must have causes quantitatively proportioned.

In other cases, Galvani makes use of *dialectical* arguments: for example, when he invokes the authority of physics upon biology in order to extend to the phenomena of the frog the physical law that opposite charges cannot exist in the same conductor. Here his arguments do not intend to prove a conclusion; rather they are directed to suggest a line of research and to convince the audience that such line indicates a program worth pursuing.

But if Galvani reasons from facts to hypothesis, in what context does he so reason? Thesis B of the *Novum Organon Reformatum* claims that a special logic of discovery does not exist and Galvani's research supports Thesis B. First, inferences I—VII are ordinary inductive inferences; and although Galvani uses analogical inferences more frequently than others, we cannot conclude that analogy — or any other form of inductive inference — is exclusive to the context of discovery. Second, the same inference can appear in different contexts, and perform different epistemic functions, as is shown by the fact that logically and historically three different sequences of inferences are possible. This is especially the case with inference VII, which is the one with the most uncertain date.

Let us first take a functional perspective. In sequence A, inference VII belongs to the original *context of discovery*. Located at the beginning of the research, it is the inference which first generates the hypothesis; its degree of strength is the value of its initial plausibility and can be used in following evaluations of the hypothesis.

In sequence B, inference VII belongs to the *context of pursuit*. Located in an intermediate state of the research where the idea of animal electricity has already been introduced, it is the inference which determines if this idea is worth developing. In this context, it works as a reinforcement of the inference of discovery, as well as the first inference of control.

Finally, in sequence C, inference VII belongs to the *context of justification*. Located at the (temporary) end of the research, it confers new evidence (important because of its independence from previous evidence) that the already introduced and then reinforced idea is a reliable hypothesis supported by facts.

If we take an historical perspective instead of a functional one, inferences I—VII turn out to be located differently. Since the hypothesis of animal electricity had already been considered by many scientists before Galvani, we could say that he was trying to test it and, therefore,

that the inferences I—VII belong to the *context of justification*. We
arrive at the same conclusion even if we take the point of view of
discoverability. We may thus conclude that the location of an inference
in one objective context rather than in another depends on the situation
in which it is made and on the objective for which it is produced, not
on the logical character of the inference itself.

Let us now look at thesis C of *Novum Organon Reformatum*. Do
Galvani's arguments have epistemic value *per se*? A consequentialist
would say "no". He would rather claim that the only inferences
endowed with epistemic value are those Galvani derived from *new*
consequences in order to test his hypothesis. Now, as a matter of fact,
a few years after the publication of the *Commentary*, during his
controversy with Volta, Galvani did in fact discover new facts that
supported his hypothesis. Arguing against Volta's theory, according to
which the contractions of the frog were due to an active power of the
metallic arc, he predicted that if nerve and muscle of the frog are put in
contact *without* any metallic arc, the frog still contracts. This is known
as Galvani's "third experiment" and it is usually considered a crucial
proof of the theory of animal electricity (Fig. 3).

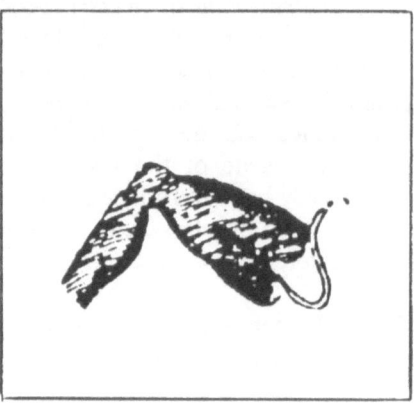

Fig. 3. Galvani's "Third Experiment" (from A. von Humboldt, *Expériences sur le
galvanisme*, Imprimerie Didot Jeune, Paris 1799).

But if we attribute epistemic value only to the third experiment, as
the consequentialist would, we do not provide an adequate account of

Galvani's research. Inferences I—VII do not have a simple intuitive value; they really prove, in their inductive and dialectical forms examined before, that the hypothesis of animal electricity is a reasonable conclusion from the known facts and laws. These inferences convinced not only Galvani but also the great majority of the scientific community including Volta, who considered the arguments of the *Commentary* as the first real experimental contribution to the theory of animal electricity. Therefore, depriving these inferences of epistemic value would mean admitting the scientific community made a mistake.

In order to avoid this unattractive conclusion, the consequentialist might reinterpret inferences I—VII. Keeping in mind that the hypothesis of animal electricity was already known, he might put these inferences into the context of justification and reconstruct them in hypothetico-deductive terms. Take, for example, inference VII which is the most suitable for this operation since it introduces a new property. The consequentialist might construe it as an inference *from* hypothesis to facts, instead of as an inference from facts *to* hypothesis. This is not implausible; in fact, Galvani's reasoning could have been as follows: if muscular fiber and the Leiden jar have the same structure, then, since the Leiden jar revives after electric discharge occuring at a certain distance, the muscle has to contract in the same way; that is:

VII*   If H: the muscular fiber and the Leiden jar have the same structure, then O: muscular fiber must have property P. O: muscular fiber has property P.
_____
H: It is probable that muscular fiber is an organic Leiden jar.

We can say that inference VII* is the consequentialist version, in the context of justification, of inference VII in the context of discovery.

However, inference VII* *does not* capture the epistemic value of inference VII. In fact, in order for VII* to have a probable conclusion, consequentialist methodology requires that O: (i) is a novel (not yet known) fact; or (ii) is different from the ones for which H was devised; or (iii) even if known, belongs to a new class of phenomena.

None of these conditions is satisfied in inference VII*. First, that muscles contracted given an electric discharge was known since Galvani's first experiment. Second, the same fact was used by Galvani to revise his hypothesis, even though it was not the only one, nor the

most important one. Finally, the fact belonged to the same class of
physiological phenomena explained by the hypothesis. As a consequence, according to the hypothetico-deductivist methodology, inference VII* has little or no epistemic value, and since inference VII has
epistemic value, inference VII* is not able to absorb it. So the
consequentialist has to conclude that the credit attributed to Galvani's
discovery before the third experiment was the result of an illusion. This
conclusion is contrary to the evidence. In the history of electro-
physiology Galvani is credited with *three* fundamental experiments, but
according to the consequentialist reconstruction, the first two wouldn't
have epistemic value.

We must admit that inference VII* is not the only possible reconstruction of VII. The following is apparently more advantageous for the
consequentialist:

VII**   If H: muscular fiber and the Leiden jar have the same
             structure, then O: the Leiden jar has property P.
             O: the Leiden jar does present property P.

             _____

             H: It is probable that the muscular fiber is an organic
             Leiden jar.

Even though this inference does not satisfy condition (*i*) because
Galvani did not anticipate O, it satisfies (*iii*) and in particular (*ii*),
because the reappearance of the electric flake in the Leiden jar
following non-proximate discharges was not part of the original
analogical basis on which Galvani built his hypothesis. But even if this
supports the consequentialist, he remains in difficulty. First of all,
reconstruction VII* is adequate only if the chronological sequence A is
valid; and this, as we saw before, is only a matter of chance. Second, in
order to make sure that condition (*ii*) is satisfied, the consequentialist
has to take into account the history prior to VII*, because only in this
way can he know if O was part of the original basis of the hypothesis.
But this means that inferences I—VII are necessary in order to confer
epistemic value on VII**. In general, this means that the logic of
discovery is an essential part of the logic of justification.

Some consequentialists agree with this conclusion, but take it in a
different sense from the one maintained here. Their thesis is: (1)
considerations belonging to the context of discovery are essential to the
logic of justification because they avoid using the same fact twice (first

in the construction, second in the valuation of the theory); however, (2) these considerations have epistemic value of zero.[24] On the contrary, the thesis defended here is that the inferences in the context of discovery are at the same level as those in other contexts because they have their own epistemic value. These inferences are not replaced by the hypothetico-deductivist ones, rather they add to them. The confirming facts have all the same importance, whether or not known to the investigators at the moment of introduction of the hypothesis, and whether or not utilized for that aim. In general, *every fact supports an hypothesis, provided a logical connection from the fact itself to the hypothesis exists*. No other temporal or heuristic consideration is relevant to the relation of support.

5. HEURISTIC, RATIONALITY, AND FALLIBILITY

Thesis (D) is all that remains to be discussed. Suppose that an inference from premises to hypothesis exists, that this inference falls into any of the many different contexts, and that it has its own epistemic value, is it possible to give some explanation about the generation of the hypothesis?

The old *Organon* relied almost exclusively on a system of rules of discovery in order to "deduce" or "induce" explanatory conclusions from phenomena. However, this point of view is defective for many reasons. First of all, there is the problem of the modalities of generation. A new idea is not the simple result of the application of rules to a set of facts; rather, it is a function of a very large class of heuristic principles.

These principles may be divided into two sub-classes. We can call the first ones *general heuristic principles*. They are principles of generation (and also of validation) that do not get their strength from the particular context in which they are used, but rather from certain general aims of scientific research. As a consequence, they are fairly stable, even if their interpretation and application are not stable or univocal. At least four classes of principles belong here:

(a) Mathematics and deductive logic.

(b) Rules of inductive logic. For example, Mill's canons are models of experimental reasoning which apply to any kind of research.

(c) Fundamental methodological rules. For instance, the rule of introducing hypotheses which accord with the known facts or the rules of refusing theories systematically at odds with facts.

(d) Universal assumptions regarding nature in general. For example, the principle of causality or the principle of proportionality of causes.

We can call the second kind of principles *contextual heuristic principles*. Their strength depends on the previous acceptance of certain scientific results. They are linked to specific programs or research traditions and, therefore, are less stable than the general principles. They are:

(e) Scientific theories with great explanatory power, for example, a theory such as Newton's imposes itself as a matrix which generates other hypotheses and exports its way of solving problems to other fields.

(f) Interpretative theories which assign phenomena to a particular ontological class and prescribe interpretations in terms of such class, for example, mechanicism or vitalism.

(g) Regulative values or images of science. Many of the themata analyzed by Holton belong to this class.

Probably classes (a)–(g) are not the only ones possible, and other classifications can be proposed; but it is easy to prove that principles of both categories are contextually operant in all scientific research.

Let us return to Galvani's case. His use of (a) is obvious. Moreover he uses (b) when he reasons analogically or relies on certain forms of eliminative induction; he uses (c) when looking for solutions in agreement with the known facts and admitted electrical theories; he uses (d) in order to exclude hypotheses whose introduction would violate certain assumptions, for example the principle of proportionality of causes. Galvani also makes use of (e) when looking for explanations in terms of mechanical actions, as well as of (f) and (g) when he commits to the view that the phenomena of the frog have an organic, non-reducible nature. From all this, it would be completely inappropriate to present Galvani's results as an inference only from phenomena; in reality, his discovery is the combined result of both phenomena and principles. More precisely, it is the result of phenomena seen in the light of those principles. The contextual presence of both kinds of heuristic principles may perhaps contribute to explaining certain typical characteristics of scientific research and to answering some traditional objections against the generativist program.

The first characteristic is the rationality of scientific discovery. If the invention of an hypothesis is an inference, then it loses that mysterious aspect it gets in the hypothetico-deductivist view. Scientific invention is not a casual mutation or a blind, lucky guess; on the contrary, it is a rational, argumentative act guided by the situation in which it takes place. It is not the same as the mechanical outcome of the elements of the situation; rather it is like an emergent property; the elements are present, what is missing is a particular combination of them. To find the proper combination is a matter of individual talent.

Another characteristic on which we may hope to shed some light is the frequent simultaneity of discoveries. If the invention of hypotheses is a guided act, it is not unlike the application of the same heuristic principles — especially the contextual ones — to the same facts and problems leads to the same or to similar results.

Other characteristics become less mysterious as well; take continuity and deep innovations, both present in the history of science.

The old inductivist *Organon* was equipped to explain continuity but not revolutions. If laws and theories are induced (or deduced) from phenomena, how can it happen that new inductions (or deductions) can so deeply upset previous ones? On the other hand, the hypotheticist *Organon* explains revolutions in a natural way, but it is not well prepared to face continuity: if laws and theories are free creations, how can they combine so harmoniously with previous ones? The hypothetico-deductivists have tried to explain this circumstance in terms of methodological restraints imposed on hypotheses in the context of justification. The new *Organon* introduces the same restraints as heuristic principles in the context of discovery. On the one hand, the stability of general heuristic principles may be used to explain the continuity of scientific research: a new theory is added to the previous one because it depends upon the same generative factors. On the other hand, the flexibility of interpretation and application of general heuristic principles, and especially the flexibility of contextual heuristic principles, may explain revolutions: a new theory can upset previous ones for it may change at least some heuristic principles of the tradition.

The new *Organon* promises to offer a model of radical or revolutionary theoretical change without the defects of holistic views. A radical or revolutionary change is one in which the new theory does not respect at least one of the deep assumptions of tradition, generally ontological (a principle of type (e) in the classification proposed above). However, because of the evident disparity between general heuristic principles and special heuristic principles, abandoning one assumption does not imply abandoning others. Even after a revolution, science is still mostly governed by its previous rules.

One final remark. As we saw before, a strong objection to the old generativist program concerned the fallibility of science: if a logic of discovery exists, how is it possible that false conclusions are derived from true factual premises and that different scientists arrive at different conclusions? While the old program didn't have appropriate answers to these questions, from the point of view of the *Novum Organon Reformatum*, there are many reasons which can be used to explain why the existence of a logic of discovery is compatible with fallibilism.

First, factual premises are never certain; and conclusions drawn from observations cannot have a strength greater than that of the premises. Second, even if we admit that a theory T inferred from O must be true for O, O does not exhaust the domain of T; it is possible that T is no longer true when extended beyond its original domain. Third, no theory T is inferred only from factual premises; we need, in addition, laws, theories, general assumptions about nature and specific assumptions about the domain of the theory; and since all these theoretical premises are questionable as much as and more than the factual ones, the conclusion is correspondingly fallible.

The last reason is perhaps the most important. As we saw before, discovery is function of a complex set of heuristic principles, among them what we called interpretative theories. Now, different interpretative theories may produce changes in the factual basis and different applications of the admitted rules.

The debate between Galvani and Volta shows very clearly how these rules, in this specific case the rules of analogy, can lead to different conclusions when the same factual basis is seen in the light of conflicting interpretative theories. The contextual heuristic principles make the logic of discovery contextual to some degree. It is a question of further philosophical analysis and historical investigation to define this degree.

Università degli Studi di Pisa   *Translated by Joseph C. Pitt*

NOTES

[1] L. Laudan. "Why Was the Logic of Discovery Abandoned?". In L. Laudan. *Science and Hypothesis*. Dordrecht: D. Reidel, 1981, pp. 181—91; see p. 191.

[2] *Ibid.*

[3] T. Nickles. "Beyond Divorce: Current Status of the Discovery Debate". *Philosophy of Science* **52** (1985): 177—206; see p. 195.

[4] *Ibid.*

[5] M. Pera. "Inductive Method and Scientific Discovery". In M. D. Grmek, R. S. Cohen and G. Cimino (eds.). *On Scientific Discovery.* Dordrecht: D. Reidel, 1980, pp. 141—65. A similar view has been maintained by R. McLaughlin. "Invention and Appraisal". In R. McLaughlin (ed.) *What? Where? When? Why?*, Dordrecht: D. Reidel, 1982, pp. 69—100; "Invention and Induction: Laudan, Simon, and the Logic of Discovery". *Philosophy of Science* **49** (1982): 198—211.

[6] *Op. cit.,* p. 188.

[7] See R. Feyman. *The Character of Physical Law.* BBC, London 1965, p. 161.

[8] T. Nickles, *op. cit.,* p. 190.

[9] P. Medawar. *Induction and Intuition in Scientific Thought.* London: Methuen, 1969, p. 29; see also p. 51.

[10] D. Hume. *A Treatise on Human Nature* (1739), ed. by L. A. Selby-Bigge. Oxford: Clarendon Press, 1967, I, III, 15, p. 175.

[11] According to Mill, a novel confirmatory fact can only "impress the uninformed"; see J. S. Mill. *A System of Logic.* 1843, 1872, New Impression 1970, London: Longman, III, XIV, 6, p. 328.

[12] I have developed this point in M. Pera. "Epistemologia evoluzionistica, cambiamento teorico e scoperta scientifica", in AA.VV. *Che cos'è pensiero? L'unità dell'estere.* Roma: Accademia Nazionale dei Lincei, 1985.

[13] K. Popper. *Objective Knowledge.* London: Clarendon Press, 1972, p. 23; see also p. 28.

[14] I have examined at length this case in my book *La rana ambigua,* Einaudi, Torino 1986.

[15] L. Galvani. *Commentary on the Effects of Electricity on Muscular Motion* (1791). Trans. by M. Glover Foley with Notes and a Critical Introduction by I. Bernard Cohen. Norwalk, Conn.: Burndy Library, 1953; see p. 47.

[16] L. Galvani. *De animali electricitate* (30 October 1786). In *Memorie ed esperimenti inediti di Luigi Galvani.* Bologna: Cappelli, 1937.

[17] See also Galvani's *Notebook* in *Memorie ed esperimenti inediti, op. cit.,* pp. 233—411; pp. 401—402.

[18] See also L. Galvani. *Electricitas Naturalis,* in *Memorie ed esperimenti inediti, op. cit.*

[19] *De consensu et differentiis inter respirationem et flammam, penicillumque electricum prodiens ex acuminato conductore Leidensis phialae de industria oneratae,* in *Opere edite e inedite del Professore Luigi Galvani,* Accademia delle Scienze dell'Istituto di Bologna, Bologna 1841.

[20] *Op. cit.,* p. 155.

[21] A. Volta. *Memoria prima sull' elettricità animale* (1791). In *Le opere di A. Volta.* Hoepli, Milano: Edizione Nazionale, 1918—29; vol. I, p. 26.

[22] *Op. cit.,* p. 75.

[23] See Galvani's *Notebook, op. cit.,* p. 406.

[24] See J. Worrall. "Scientific Discovery and Theory-Confirmation". In J. Pitt (ed.) *Change and Progress in Modern Science.* Dordrecht: D. Reidel, 1985, pp. 301—331; E. Zahar. "Logic of Discovery or Psychology of Invention?" *British Journal for the Philosophy of Science* **34** (1983): 243—61.

RACHEL LAUDAN

THE RATIONALITY OF ENTERTAINMENT
AND PURSUIT*

INTRODUCTION

Science has traditionally been taken to be the paradigm of rationality. If any beliefs are rationally warranted, they are beliefs in the best scientific theories of the day. Furthermore, the rationality of science has been taken to reside in the context of justification. Since one of the most striking features of science is the rapidity with which scientists change their beliefs, the question of rationality has usually been formulated as one of understanding the circumstances in which it is rational for a scientist to abandon one belief for another.[1]

Through most of this century philosophers of science relegated discovery to a realm where questions about the rationality of the agent were irrelevant. Any method of making discoveries, rational or irrational was acceptable. It was the way the discoveries were subsequently tested that mattered. In the last decade or so this traditional account of rationality has been gradually extended.[2] The historical turn in the philosophy of science has forced attention to the process rather than the products of the scientific enterprise. Philosophers have realized that if rationality is restricted to a consideration of the circumstances under which theories can be accepted as true then large portions of the scientific enterprise fall outside their ken. As a result, instead of trying to produce an account of rationality limited to the context of justification or acceptance, philosophers have been considering when other attitudes towards scientific theories might be regarded as rational. Suggestions for other epistemic attitudes that deserve study include preparation, discovery, entertainment, and pursuit as well as acceptance. In this paper I shall examine two of those possibilities — entertainment and pursuit — in the light of a particular case study, namely the shifting attitudes of scientists towards continental drift theory during the first half of this century.

MODALITIES OF APPRAISAL

Four different modalities of appraisal capture scientists' epistemic

203

J. C. Pitt and M. Pera (eds.), Rational Changes in Science, 203—220.
© 1987 by D. Reidel Publishing Company.

attitudes to drift (or, for that matter, to other theories). They are
rejection, entertainment, pursuit and acceptance. I shall say little about
acceptance and rejection, since they are so fully discussed in the
philosophical literature, and instead I shall concentrate on entertainment and pursuit. By entertainment I mean the willingness of a scientist
(or the scientific community) to admit a new theory to the group of
theories worthy of consideration. By pursuit I mean the the decision of
a scientist (or group of scientists) to develop a theory, whether by
articulating its theoretical base or by collecting evidence for its
appraisal.[3]

In order to decide whether it is rational to entertain or to pursue a
theory, we have first to define rationality. For the purposes of this
paper the common, though not uncontroversial, definition of rationality
as an agent's choice of actions that maximize the achievement of his
ends (instrumental rationality) will serve. Of course, this definition is
subject to the criticism that it says nothing about the rationality of the
aims in question. But since all the scientists in the debate over drift
shared a common aim (to develop theories that explained the origin of
the major features of the earth's crust), we can safely ignore the
criticism in this case. Their disagreements were about how best to
achieve that end, that is to say, about how to decide between the
different theories that had put forward to explain crustal features.

Rationality, on this definition, involves selecting the best available
means for achieving an end. Hence, it may well be the case that
adopting a particular means is irrational, not because it fails to achieve
that end, but because it is not the *best* means to achieve that end. Thus
examining the rationality of adopting a certain epistemic attitude to a
theory necessarily involves examining the theory's credentials in a
comparative context.

THE CHOICE OF CASE STUDY

During the mid-1960s, almost all geologists came to accept a version of
the theory of continental drift. This is one of the most dramatic cases of
scientific theory change in the last thirty years.[4] Geologists had been
familiar with the theory for for half a century before they accepted it.
Alfred Wegener had proposed it in 1915, as a solution to the classic
geological problem of explaining of the major features of the earth's
crust, the continents and oceans, the mountain chains and oceanic

islands. Since the explanation that had reigned since the beginning of the nineteenth century — the theory that these features had been caused by the cooling and contraction of an originally hot globe while the continents and oceans had maintained essentially the same positions — was in trouble, drift was widely discussed in the mid 1920s. Even so, for fifty years hardly any earth scientists accepted the theory, at least in North America, and, by the 1950s, most dismissed it with some scorn. Yet a decade later, they switched *en masse* to the theories of sea-floor spreading and plate tectonics that, like drift, asserted that mountain chains and oceanic islands were caused by the movement of large portions of the earth's crust, and that continents and oceans had dramatically shifted their positions on the earth's surface. Such a complete and rapid change of belief clearly challenges us to determine whether it was rational, or, as has sometimes been asserted, simply a bandwagon effect. As such it merits attention by philosophers interested in scientific rationality.

In an attempt to understand this sudden reversal of belief, many commentators, especially those without a background in geology, have resorted to one of the two following hypotheses: either they have suggested that Wegener's drift theory must have been pseudo-scientific, or, perhaps more plausibly, they have suggested that scientists' attitudes towards Wegener's theory must have been unscientific, and their later conversion irrational. These alternatives are hard to avoid if a simple dichotomy of acceptance and rejection is taken to exhaust the range of epistemically reputable attitudes to scientific theories. But if it can be shown that there are conditions under which it is rational to entertain or pursue a new theory without accepting it as true, and if it can be further shown that those conditions were satisfied during the early history of drift theory, then the dilemma can be avoided.

In this paper, I shall suggest some plausible conditions for rational entertainment and pursuit, and use them to interpret the actions of geologists from the 1920s to the 1960s. I shall argue that they were rational to entertain the theory, and that they were equally rational to neither pursue nor accept it before the mid-1950s. Because it was testable in principle, they regarded it as scientific not to be rejected out hand. Because rigorous tests for it were readily at had, they decided they had to suspend judgement about its acceptability. Because testing demands a commitment of limited resources of time, money and skill,

the decision about whether to pursue the theory hinged on an assessment of whether such a commitment would produce the necessary
results, and produce them more quickly or at less cost than the pursuit
of competing theories. The question facing scientists during drift's early
history, as in so many other cases, was not whether the theory was
testable and hence scientific, nor whether it had passed enough tests to
be accepted, but whether it should be entertained and pursued. They
did not reject it. But neither did they think it the best theory to pursue
in order to achieve their epistemic ends. As we shall see, the dramatic
change of attitude to drift in the early 1960s resulted from the
discovery of compelling new evidence from completely unexpected
sources.

Examining the history of continental drift provides a nice test for
intuitions about the criteria that scientists bring to bear when they
adopt a particular epistemic attitude to a theory. No one criterion, nor
any set of criteria will guarantee that scientists can unerringly pick out
the theory that will ultimately turn out to be successful. Indeed the
criteria did not pick out drift although turned out to be successful.
Nonetheless over the long run using some rational criteria ought to pick
out more successful theories than random selection. Since, as I have
argued, the appropriateness of any cognitive attitude is context dependent I shall begin by describing the theoretical situation in geology
when Wegener first introduced continental drift.

THE GEOLOGICAL BACKGROUND

Continental drift was just one among many theories about the origin of
continents and oceans put forward in the first decades of this century
(Greene, 1982). The flurry of theorizing was prompted by the collapse
of the traditional explanation for the major features of the earth's crust.
Theories of a cooling, contracting earth had ruled throughout the
nineteenth century, and were consistent with physical and cosmological
theory, intuitively consistent with the geological record, and without
serious long-term rivals. Eduard Suess's monumental four-volume
work, *The Face of the Earth*, published between 1885 and 1909
marked their zenith (Suess, 1904—1909). His version of the theory was
supported by a wealth of detailed evidence, and was held in such high
esteem that, on the appearance of the final volume, Suess received
telegrams of congratulation signed by the entire membership of the

leading geological societies of Great Britain and the United States of America.

But at the very time that Suess was writing, the contraction theory was running into serious trouble. Detailed study of the Alps revealed that the contraction of the earth's crust in the Tertiary (when these mountains were being elevated) was of the order of tens or even hundreds of miles, far more than could be explained on even the most dramatic theory of earth contraction. Simultaneously the discovery of radioactivity, and the consequent recognition of a heat source within the earth, necessitated a re-thinking of the theoretical base of contraction theory. When combined with other disquieting lines of evidence that need not concern us here, geologists found that the framework on which they had hung so many of their specific geological theories was in a state of collapse.

This led geologists to formulate and entertain a spectrum of new theories that might not have seen the light of day had there been stronger competition. Some, such as Harold Jeffreys in England and Hans Stille in Germany attempted to re-vamp contraction theory in the light of the criticisms (Jeffreys, 1924; Chamberlin, 1907). Others tried more radical solutions, including John Joly's theory based on radio-activity, Thomas Chamberlin's planetismal hypothesis, or Bailey Willis's eclectic theory (Willis, 1907; Chamberlin, 1907; Joly, 1909). And then there was continental drift theory.[5]

WEGENER'S ALTERNATIVE TO CONTRACTION THEORY

Wegener accepted Suess's description of the structure of the earth, including the hypothesis of former mega-continents (Wegener, 1915—36 and 1966). But he argued that our present ocean basins and mountain chains are better explained as the result of the break-up of the mega-continents, and the steady drift of the pieces (our present continents) away from the poles than as collapse and contraction features. Oceans were basins in which the lighter continents floated, and mountains were pushed up by resistance as the continents moved showly around the surface of the globe.

As independent evidence for his theory, Wegener cited the apparent fit of the two coastlines of the Atlantic, the similarities of rocks and fossils across certain oceans, and the apparent evidence for a Permo-Carboniferous glaciation in locations that were implausibly close to the

Equator unless the continents had moved. He freely admitted that the
nature of the force supposedly causing the continents to move was a
problem. He did not have much confidence in either of his own
proposals, namely, the so-called "pole-fleeing" force, or the force
exerted on the crust by the tidal forces between the earth and the
moon. But he argued that establishing that drift *had* occurred was more
urgent than establishing *how* it had occurred. Wegener embarked on a
series of geodetic measurements in the hope that they would give
conclusive proof that the continents were drifting. In 1930 he died in
the attempt on the Greenland icecap.

THE EARLY REFUSAL TO ACCEPT DRIFT

There is a popular myth that continental drift was dismissed immediately by the scientific community as pseudo-scientific, and thereafter
ignored.[6] This is far from the truth. The theory was seriously considered by many of the leading geologists of the day, particularly after
the appearance of English, Russian, French, Spanish and Swedish
translations of the third (1922) edition in the mid-1920s. The reception
varied somewhat from country to country, and I shall concentrate on
Britain and America. Geologists discussed drift at the 1922 meetings
of the British Association for the Advancement of Science, at the
Geological Society of America, at a special meeting of the Washington
Academy of Sciences in 1923, and at an invitational meeting of the
American Association of Petroleum Geologists in 1926 (Lake, 1922
and 1923; Waterschoot van der Gracht, 1928; see also Marvin, 1973).
    The latter is particularly revealing because of the expertise of the
fourteen scientists involved. Wegener was there, as well as an American, Frank Bursley Taylor, who had put forward a similar theory, and
one of their strongest supporters, Waterschoot van der Gracht. Rollin
Chamberlin (Thomas Chamberlin's son and, on these issues, his disciple),
Bailey Willis, and John Joly, represented alternatives to drift theory.
The remainder of the participants were leading practitioners from
different geological subfields. With the exception of Wegener, Taylor
and Waterschoot van der Gracht, the scientists at the meeting were not
prepared to accept Wegener's theory as true. They thought its evidential support inadequate and the force that was supposed to move the
continents mysterious.
    Taking the evidential support first, geologists found drift theory

unimpressive. The consensus was that Wegener's data were too ambiguous to justify acceptance of his theory. It was not clear, for example, whether the two sides of the Atlantic should fit along the present coastlines, or along the edge of the continental shelf. Indeed since the theory asserted that mountain chains were caused by the crumpling of the leading edge of the moving continents, it was not clear that any fit at all should be expected. Some fossil evidence supported the theory; some did not, and in any case geologists had an alternative — if problematic — explanation in terms of former land bridges. Paleoclimates were notoriously difficult to reconstruct, and evidence of the Permo-Carboniferous glaciation was by no means universally accepted. And with the exception of Wegener, everyone believed with good reason that the available geodetic techniques were too insensitive to detect the relative movement of the continents. Given all this, Wegener's failure to produce a plausible mechanism for the movement of continents became the focus of many objections to the theory. Not even his greatest admirers accepted them.

The general attitude of geologists was one of regret. If only Wegener's theory had had more evidence in its favor it would have been a welcome solution to the problem of the origin of continents and oceans.[7] As it was, while agreeing about the scientific status and potential importance of the theory, almost no American or British scientists accepted it in the 1920s. But very few rejected it either.

THE ENTERTAINMENT OF DRIFT BY THE SCIENTIFIC COMMUNITY

Most geologists entertained Wegener's theory, without undertaking the further commitment of pursuing it, and certainly without going so far as to accept it. Were they rational to entertain it? I shall argue that they were since it satisfied four plausible conditions for rational entertainment, namely centrality, timeliness, preparedness and testability. The problems that Wegener's theory was intended to solve was widely agreed to be central to geology (centrality). He put forward the theory when there was a widely perceived need for such a solution (timeliness). The solution Wegener proposed demonstrated his mastery of the data and techniques of geology (preparedness). Finally his theory was, in principle at least, testable (testability). Let me discuss these criteria of rational entertainment in turn.

A new theory that addresses an important problem has a higher claim on scientists' attention than one that addresses a fringe problem. Admittedly fringe and center are difficult to define, so I shall simply accept scientists' judgement on the matter. And admittedly fringe problems can, and in the history of science often have, turned out in retrospect to be important stimuli to the development of major new theories. One need only think, for example, of the significance that certain seemingly trivial and anomalous refraction phenomena had in the development of optics. But a scientist who attempts to persuade his peers to entertain a major new theory developed to account for a fringe problem faces a harder task than the scientist who attempts to persuade them to entertain one developed to account for a central problem. To show that a fringe problem demands the overthrow of well-entrenched beliefs requires that the scientist in question formulate an alternative to these beliefs, or at least a theory that shows why problems formerly thought to be central should be relegated to the fringe. That is a tall order. Scientists who seek acceptable theories within a given domain are surely warranted attending most carefully to those new theories that address the major problems of the discipline. On such grounds Wegener's theory scored high. Geologists did regard the problem of the origin of continents and oceans as central to geology.

Scientists are also rational to entertain timely theories, that is to say, theories advanced when formerly successful theories run into trouble. Again this criterion does not uniquely pick out theories that will be successful. New theories may be developed when theories in a given domain remain successful, and yet over time the new theories may prove superior. Nonetheless, scientists would be less than reasonable to spend much time entertaining alternatives to highly successful theories, or to fail to consider alternatives to unsuccessful old theories.

In addition, scientists are reasonable to entertain theories put forward by a well-prepared scientist who knows the state of the discipline. Wegener himself, although originally a meteorologist, showed an enviable mastery of the state of a wide range of geological specialities. He did not regard the overthrow of the Suessian theory as reason to begin from scratch, but incorporated much of it into his own theory. As a result, at least in America, he was widely criticized as being almost too Suessian. I do not want to suggest that Wegener, in attempting to preserve as much as possible of Suess's theory, was hewing to any principle of cumulation or to the belief that any acceptable successor theory must incorporate all the true consequences of the old. His theory patently did not do so. Rather, I am making the more modest claim that in a situation like the one that faced geologists in the early twentieth century, new theories that incorporate well-established results are more likely to be entertained than new theories that do not.

Finally, and this hardly needs saying, scientists are reasonable to entertain testable theories. To entertain a theory so vague or so ill-formulated that it was without testable consequences would fly in the face of all our intuitions about the empirical nature of science. Testability has long been a criterion for scientific status of a theory. Once again Wegener's theory satisfied this criterion, although as we shall see, only just.

These criteria may not be either singly necessary or jointly sufficient to determine the rationality of entertaining a new theory. They may not exhaust the criteria that could be put forward. But I want to suggest that they are all plausible conditions for entertainment. Since Wegener's theory satisfied them, scientists were rational to entertain it in the 1920s.

EARLY REJECTION OF CONTINENTAL DRIFT THEORY

Before going on to consider subsequent attitudes to drift, we should note that two conference participants rejected the theory out of hand. They were not typical of the general mood of the meeting, although some historians and philosophers have made it seem so.[8] Therefore their reasons for rejection need to be examined.

Edward Berry, Professor of Paleontology at Johns Hopkins, argued that drift theory (and by implication any other theory of the origin of the major features of the crust of the earth for that matter) was premature and unscientific simply because it was a hypothesis. Without more supporting facts, Berry refused to entertain the hypothesis. But he had a much more stringent view of the tests a theory should pass when first put forward than most of his contemporaries, or most modern scientists. (Waterschoot van der Gracht, 1928; 194—196). Vanishingly few theories would survive such requirements. By contrast, Rollin Chamberlin, Professor of Geology at the University of Chicago, feared that scientists in other disciplines might think geology unscientific if they realized that there was so much debate about theoretical issues. He argued that geologists should adopt some one theory (presumably his own) and stick to it — an ironic, if understandable stance for the son of the geologist who had argued that the "method of multiple working hypotheses" was the appropriate methodology for geology (Waterschoot van der Gracht, 1928; 83—87). Therefore although it is historically and sociologically interesting that Berry and Chamberlin claimed that drift was unscientific — the ultimate dismissal after all — they did not do so on grounds that most of their contemporaries or successors would accept. Indeed, it was not drift *per se*, that they were criticizing, but the state of the discipline.

CONTINENTAL DRIFT THEORY AND PURSUIT

Few scientists at the AAPS meeting, apart from Berry and Chamberlin, quarrelled with Waterschoot van der Gracht's conclusion that the drift hypothesis opened "a vista for acceptable explanation of a series of geological problems, which so far had received no adequate answer" (Waterschoot van der Gracht; 75). But few of them left the conference intending to pursue the theory. We have to consider what would have been required to convince scientists to change from somewhat passively entertaining a theory to actively pursuing it. Wegener was scarcely likely to persuade scientists like Chamberlin, Joly, and Willis, who had formulated their own solutions to the problem of the origin of continents and oceans to spend time exploring the ramifications of a rival. But given the significance of the problem, and the credit that would rebound to anyone who could make the solution compelling, one might have expected a number of geologists to further articulate and test drift. More generally, why is a theory that successfully meets the criteria for entertainment not also pursued?

The conditions that have to be satisfied for rational pursuit resemble those for rational entertainment. Centrality, timeliness, preparedness and testability remain important. But their weighting differs. Preparedness becomes less important and testability more important. And there is a further condition, namely that the theory being considered show more rapid improvement than its rivals (L. Laudan, 1977, 108—110). Drift was not to show such improvement for many years.

The comparative character of theory appraisal is more important for pursuit than for entertainment. Pursuit requires a much greater commitment of resources than entertainment. Thus the theory has to appear a more promising subject of pursuit than its competitors. Wegener's theory was competitive with respect to centrality and timeliness. It fell down rather badly on testability.

Before a scientist (or a scientific community) invests a great deal of limited research time and money in testing a theory, he wants to assure himself that there is a reasonable chance of getting definitive results from those tests. To choose an intractable area of research is to risk frustration and obscurity, as well as to squander precious resources. In the case of Wegener's theory, geologists were very skeptical about the the feasibility of further tests in the 1920s. They believed the geodetic measurements needed to detect continental movement were unobtainable with the instruments then available. A more promising approach was to try to show more conclusively than Wegener had done that the coastlines, fossils and paleoclimates on (say) the two sides of the Atlantic did in fact match as Wegener's theory predicted. But patently, individul field surveys on such global issues of correlation would never provide enough data to be decisive. Without numbers of new detailed co-ordinated regional studies, nothing was likely to emerge that Wegener and others had not already been able to piece together from a survey of the literature. In addition, most of the best potential evidence came from the southern hemisphere, thus putting it quite literally out of the range of American and British geologists.

Parenthetically, it is worth pointing out here that the limited commitment of resources required for entertainment means that substantial numbers of a community of scientists can entertain a new theory. Pursuit is much more likely to be undertaken by an individual or a small number, working at a time when the bulk of their peers are still entertaining the theory. Acceptance, like entertainment, tends to be more community wide, and for similar reasons.

The small number of geologists who pursued drift by and large decided that the theoretical problems of drift theory were more likely to be solved by an individual scientist than the empirical problems we have just mentioned. Thus several of them, including influential figures like Reginald Daly and Arthur Holmes, tried to find an adequate cause for the movement of the continents. They put forward a variety of hypotheses including continental sliding and convection currents (Daly, 1926; Holmes, 1931; Chelikowsky, 1944; see also Frankel, 1978). Most geologists regarded these sketchy hypotheses as intriguing rather

than compelling. They wanted evidence that drift had occurred, not
simply that it could have occurred.

In short, few scientists elected to pursue drift. Compared to the rival
hypotheses, it was at least as difficult, if not more difficult, to test and it
proved resistant to further conceptual articulation.

DRIFT 1930—33

By the mid-1940s, thirty years after Wegener had first proposed drift
theory and twenty years since his theory had become known in the
English-speaking world, most American and British geologists were
more, not less, skeptical about the theory than they had been initially.
The few scientists who had pursued the evidence for the theory had
found negative results. G. G. Simpson's 1943 study of fossil mammals
was the most famous case. Fossil mammals had been a favorite choice
for testing global claims since Cuvier's time because they are large and
relatively easy to identify. Ostensibly testing three different models of
the origins of continents and oceans, Simpson was particularly hostile
to drift. Allowing that mammals on different continents showed substantial similarities that could only be explained on the assumption that
the continents had once been joined, Simpson argued that drift offered
no explanation of these similarities. Drift could explain the making or
breaking of continental contact at only one point in time. What was
needed to explain the fossil record was a series of such events. He
therefore opted for permanent continents with oscillating land bridges,
and concluded that the mammal record offered no support for drift.
The Wegnerian, DuToit (DuToit, 1944), offered an impassioned re-
joinder and Chester Longwell (Longwell, 1944) a more temperate
sequel. Both urged that drift be kept in the group of theories under
consideration, since fossil mammals were only one among many
potential lines of evidence. But most geologists were swayed by
Simpson's argument. They could scarcely be otherwise, given that there
was no new positive evidence to counteract it. The theory had not been
refuted, but it was in a state of stasis, or even decline. As such it was
unlikely to attract further pursuit.

In short, far from showing a greater rate of progress than rival
theories, drift stood still, or even regressed between 1930 and 1955.
When compared with efforts, say, by Jeffreys to revivify contraction
theory, drift made a very poor showing. Some even went so far as to

say it should be relegated to the status of "ein Märchen" — a fairy tale (Willis, 1944).[9] In the circumstances, scientists were rational not to devote further effort to pursuing drift. Until the late 1950s, nothing happened to change scientists' attitudes.

THE ACCEPTANCE OF DRIFT 1955—1970

By 1960, drift was once again a live option, and by 1970 a modified form of the theory was almost universally accepted in the geological community. Far from being a fad, fashion, or mere bandwagon effect, geologists responded to some unexpected, but compelling new evidence. New techniques for oceanographic exploration and for the investigation of the history of the earth's magnetic field had been developed after the war. The results scientists obtained by their use re-opened the whole question of drift.

In the course of testing rival theories about the origin of the earth's magnetic field, physicists had developed a new and highly sensitive magnetometer. Modifying this instrument to measure remanent magnetism — the fossil magnetism preserved in certain rocks from the time of their original formation — they discovered two very surprising results. First, the angle of inclination of the earth's magnetic pole relative to geographical north appeared to have shifted over time. Second, the direction of the earth's magnetic field had reversed or "flipped" on a number of occasions in the past. The record of these reversals provided a potential new time scale which was soon to prove crucial in the acceptance of drift. However, it was the discovery of the field's change of inclination that triggered the reconsideration of drift theory. After systematically testing and eliminating a variety of ways of accounting for this phenomenon, a number of geophysicists became convinced that, in order to explain the magnitude of the effect, the continents must have moved. Furthermore, the timing and direction of their movement was very similar to that postulated by Wegener (Runcorn, 1962; Blackett et. al., 1965). The importance of the drift phase is driven home by the fact that there was a ready-formulated theory to which they could turn.

By itself, the directional paleomagnetic evidence was not enough to persuade geologists to accept drift. It demanded such technical mastery and was so riddled with theoretical assumptions that many found it suspect. But it did raise again the possibility of pursuing drift. Traditional lines of evidence, such as the fit on opposite sides of the Atlantic were taken out, dusted off, and re-examined, but found to be no more convincing for being run through a computer than they had been when tested simply by sliding sheets of transparent plastic over the surface of a globe.

A combination of oceanographic and geomagnetic reversal data finally clinched the acceptance of drift. Oceanographic exploration had revealed a number of puzzling features about the ocean floors, in particular the existence of massive mid-ocean "ridges" with peculiar geophysical properties. A plethora of theories were put forward to explain these ridges. Among them was Harry Hess's suggestion that they marked the spots where upwelling convection currents in the mantle were solidifying to produce new crust which was carried apart by the continuing up flow of material. If the ocean floors moved apart, then the continents at their edges presumably moved too. Confirmation for this "sea-floor spreading" hypothesis came from the record of geomagnetic reversals. Vine and Matthews suggested that the odd magnetic patterns found around the mid-ocean ridges might be the lateral record of these reversals, frozen in rock as the sea floor formed and moved away from the ridges. To everyone's surprise, including their own, this conjecture was confirmed with a precision that left little room for doubt that the continents moved (Vine and Matthews, 1963). Shortly afterward, the new theory of plate tectonics, which had some significant differences from the older drift hypothesis, but retained the idea of continental movement, was proposed, tested and confirmed. By no later than 1970, it had become the basic organizing principle in geology.

CONCLUSION

Despite recent descriptions of continental drift as fringe science, borderline science, unconventional science or even pseudo-science, geologists always regarded it, and rightly so, as a scientific theory — badly wrong, maybe, but scientific. Drift satisfied the conditions for the rational entertainment of theories. But only a few scientists believed it satisfied the conditions for rational pursuit. Let me review the evidence presented here with these issues in mind.

As we have seen, geologists entertained drift since it was a testable theory that addressed a central and timely geological problem. Furthermore this theory was not tangential to the mainstream interests of the discipline (as in the case of Velikovsky) nor did it postulate anomalous entities (as in the case of the Loch Ness monster or meteorites). Rather it fell squarely within the discipline of geology. This crucial aspect of drift theory has often been overlooked by those who have assumed that Wegener was trying to prove the previously-unnoticed "fit" of the continents, and thus drawing it to the attention of geologists for the first time.[10] If this had been all that Wegener was trying to do, geologists would have been quite right to ignore the theory. But this history of Wegener's aims is misleading. His theory was much more ambitious than an account of apparent continental fit, and geologists were right to entertain it.

Most scientists did not pursue the theory. Even given the likelihood, already referred to, that the number of scientists pursuing a theory is likely to be at best a small subset of the number of scientists entertaining it, drift theory still did not attract many pursuers. Scientists did not collect evidence to test it thoroughly. Although drift satisfied some criteria for pursuit — centrality and timeliness, for instance, — it was, in practice, hard to test, and showed little progress compared to its rivals. There was no straightforward way that an individual scientist, or even a group of them, could amass compelling evidence for (or against) drift, given the available techniques. Thus after an initially favorable reception, drift was relegated to a scientific limbo where for thirty years the theory was kept alive. Its failure to gain ground becomes another argument against the theory.

Scientists' attitudes to drift would probably not have changed had new oceanographic and paleomagnetic techniques not been developed. They were designed with quite other purposes in mind, but, by sheer luck, provided other theories of the origin of continents and oceans. Once drift theory became relatively easily testable, scientists pursued it with vigor. As new evidence accumulated, its adherents modified it to fit. Drift began to show significant progress relative to its rivals. In a dramatically short period between 1960 and 1970, almost all earth scientists came to accept a modified version of drift as the best means to their end of understanding the major features of the earth's surface.

Virginia Tech

NOTES

* This paper was written in 1982/83. Although I have not changed my mind about the major argument of the paper, if I were writing it today I would include references to the substantial body of secondary literature on drift that has appeared since then.
¹ See Newton-Smith (1981) for a survey of the classic positions.
² A useful summary of the shift is given by Thomas Nickles in the introduction to his (1980).
³ I owe this term to Larry Laudan (1977).
⁴ Useful histories of drift can be found in Marvin (1973); Hallam (1973); Takeuchi (1967); Uyeda (1978); and Frankel (1976); (1978); (1979).
⁵ See also the comment of the President of the Royal Geographical Society, "Some theory of this kind is required to explain facts which have long been known to geologists" (Lake, 1923).
⁶ The Radners' suggestion that Wegener's theory is explanation-by-scenario without reference to general laws, and thus dangerously close to Velikovsky's theory is misleading (Radner and Radner, 1982: 91). Much of historical science, whether it be cosmology, geology, paleontology or embryology, does appeal to laws.
⁷ See the comment of the President of the Royal Geographical Society. "The impression left on my mind by the discussion is that geologists, as a whole, regret profoundly that Professor Wegener's theory cannot be proved to be correct" (Lake, 1923 quoted in Marvin, 1973; 86).
⁸ See Radner & Radner (1982).
⁹ Ron Westrum, in a recent issue of the Social Psychology of Science Newsletter, characterized this remark as a scientific "blooper." Yet there was considerable justification for Bailey Willis' skepticism.
¹⁰ It is true that Wegener claims that "The first concept of continental drift came to me as far back as 1910, when considering the map of the world, under the direct impression produced by the congruence of coastlines on either side of the Atlantic" (Wegner, 1966, 1). However, this is the first idea for the mechanism of the origin of continents and oceans. Had the Suessian theory still been regarded as adequate, Wegener would not have been searching for such a solution.

BIBLIOGRAPHY

Chamberlin, T. C. and Rollin D. Salisbury. (1907). *Geology*, 2nd edn. (New York: Holt).
Blackett, P. M. S., Sir Edward Bullard and S. K. Runcorn. (1965). "A Symposium on Continental Drift," *Philosophical Transactions of the Royal Society of London Series A* **258**.
Chelikowsky, J. R. (1944). "A New Idea on Continental Drift." *American Journal of Science* **242**: 673—685.
Cox, A. (1973). *Plate Tectonics and Geomagnetic Reversals*. San Francisco: Freeman.
Daly, R. A. (1926). *Our Mobile Earth*. New York: Charles Scribner.

Du Toit, A. L. (1944). "Tertiary Mammals and Continental Drift." *American Journal of Science* **242**: 145—63.

Du Toit, A. L. (1937). *Our Wandering Continents*. Edinburgh: Oliver and Boyd.

Frankel, H. (1976). "Alfred Wegener and the Specialists." *Centaurus* **20**: (305—324).

Frankel, H. (1978). "Arthur Holmes and Continental Drift." *British Journal for the History of Science* **11**: 130—150.

Frankel, H. (1979). "The Reception and Acceptance of Continental Drift Theory as a Rational Episode in the History of Science." In *The Reception of Unconventional Science*. Edited by S. H. Mauskopf. Washington: AAAS. Pp. 51—89.

Greene, M. (1982). *Geology in the Nineteenth Century: Changing Views of a Changing World*. Ithaca: Cornell University Press.

Hallam, A. (1973). *A Revolution in the Earth Sciences: From Continental Drift to Plate Tectonics*. Oxford: Clarendon Press.

Hess, H. H. (1962). "History of Ocean Basins." In A. E. J. Engel, H. L. James and B. F. Leonard, eds. *Petrologic Studies: A Volume to Honor A. F. Buddington*. New York: Geological Society of America.

Holmes, A. (1931 for 1928—9). "Radioactivity and Earth Movements." *Transactions of the Geological Society of Glasgow* **27**: 567—606.

Jeffreys, H. (1924). *The Earth, Its Origin, History, and Physical Constitution*. Cambridge: The University Press.

Joly, J. (1909). *Radioactivity and Geology*. London: Constable.

Lake, P. (1922). "Wegener's Displacement Theory." *Geological Magazine* **59**: 338—46.

Lake, P. (1923). "Wegener's Hypothesis of Continental Drift." *Geographical Journal* **61**: 179—94.

Laudan, L. L. (1977). *Progress and Its Problems: Towards a Theory of Scientific Growth*. Berkeley: University of California Press.

Laudan, R. (1981). "The Method of Multiple Working Hypotheses and the Origin of Plate Tectonic Theory." In T. Nickles (ed.) *Scientific Discovery: Case Studies*. Dordrecht: D. Reidel. Pp. 333—343.

Longwell, C. R. (1944). "Some Thoughts on the Evidence for Continental Drift," *American Journal of Science* **242**: 218—231.

Longwell, C. R. (1944). "Further Discussion of Continental Drift." *American Journal of Science* **242**: 514—515.

Marvin, U. B. (1973). *Continental Drift: The Evolution of a Concept*. Washington, D.C.: Smithsonian Institution Press.

Newton-Smith, W. H. (1981). *The Rationality of Science*. London: Routledge and Kegan Paul.

Nickles, T. (ed). (1980). *Scientific Discovery, Logic, and Rationality*. Dordrecht: D. Reidel.

Poldervaart, A., ed. (1955). "Crust of the Earth." *Geological Society of America Special Paper*, 62. P. 5.

Radner, Daisie and Michael Radner. (1982). *Science and Unreason*. Belmont, California: Wadsworth.

Runcorn, S. K. ed. (1962). *Continental Drift*. New York and London: Academic.

Simpson, G. G. (1943). "Mammals and the Nature of Continents." *American Journal of Science* **241**: 1—31.

Suess, E. (1904—1909). *Das Antlitz der Erde* [1885—1909]. Translated into English by Hertha B. C. Sollas under the direction of W. J. Sollas. 4 vols. Oxford: Clarendon.

Takeuchi, H. et al. (1967). *Debate about the Earth: Approach to Geophysics through Analysis of Continental Drift*. San Francisco: Freeman.

Taylor, F. B. (1971). "Bearing of the Tertiary Mountain Belt on the Origin of the Earth's Plan." *Geological Society of America Bulletin* 21: 179–226.

Uyeda, S. (1978). *The New View of the Earth*. San Francisco: Freeman.

Van Waterschoot ̇van der Gracht, W. A. J. M., et al. (1928). *Theory of Continental Drift: A Symposium*. Tulsa: American Association of Petroleum Geologists.

Vine, F. and D. Matthews. (1963). "Magnetic Anomalies over Oceanic Crust." *Nature* 119: 947–949.

Wegener, A. (1915—36). *Die Entstehung der Kontinente und Ozeane*. Braunschweig: Friedrich Vieweg. 1st edition, 1915; 2nd edition, 1920; 3rd edition, 1922; 4th edition, 1924; 4th edition revised, 1929; 5th edition, revised by Kurt Wegener, 1936.

Wegener, A. (1966). *The Origin of Continents and Oceans*, translated from the 3rd (1922) German edition by J. G. A. Skerl. London: Methuen, 1926. Translated from the 4th (1929) German edition by John Biram. New York: Dover.

Willis, b. (1907). "A Theory of Continental Structure Applied to North America." *Geological Society of America Bulletin* 18: 389–412.

Willis, B. (1944). "Continental Drift, ein Märchen." *American Journal of Science* 242: 509–513.

Wilson, J. T. (1965). "A New Class of Faults and Their Bearing on Continental Drift." *Nature* 207: 343–347.

INDEX OF NAMES

BOSTON STUDIES IN THE PHILOSOPHY OF SCIENCE

Editors:

ROBERT S. COHEN and MARX W. WARTOFSKY

(Boston University)

1. Marx W. Wartofsky (ed.), *Proceedings of the Boston Colloquium for the Philosophy of Science 1961–1962.* 1963.
2. Robert S. Cohen and Marx W. Wartofsky (eds.), *In Honor of Philipp Frank.* 1965.
3. Robert S. Cohen and Marx W. Wartofsky (eds.), *Proceedings of the Boston Colloquium for the Philosophy of Science 1964–1966. In Memory of Norwood Russell Hanson.* 1967.
4. Robert S. Cohen and Marx W. Wartofsky (eds.), *Proceedings of the Boston Colloquium for the Philosophy of Science 1966–1968.* 1969.
5. Robert S. Cohen and Marx W. Wartofsky (eds.), *Proceedings of the Boston Colloquium for the Philosophy of Science 1966–1968.* 1969.
6. Robert S. Cohen and Raymond J. Seeger (eds.), *Ernst Mach: Physicist and Philosopher.* 1970.
7. Milic Capek, *Bergson and Modern Physics.* 1971.
8. Roger C. Buck and Robert S. Cohen (eds.), *PSA 1970. In Memory of Rudolf Carnap.* 1971.
9. A. A. Zinov'ev, *Foundations of the Logical Theory of Scientific Knowledge (Complex Logic).* (Revised and enlarged English edition with an appendix by G. A. Smirnov, E. A. Sidorenka, A. M. Fedina, and L. A. Bobrova.) 1973.
10. Ladislav Tondl, *Scientific Procedures.* 1973.
11. R. J. Seeger and Robert S. Cohen (eds.), *Philosophical Foundations of Science.* 1974.
12. Adolf Grünbaum, *Philosophical Problems of Space and Time.* (Second, enlarged edition.) 1973.
13. Robert S. Cohen and Marx W. Wartofsky (eds.), *Logical and Epistemological Studies in Contemporary Physics.* 1973.
14. Robert S. Cohen and Marx W. Wartofsky (eds.), *Methodological and Historical Essays in the Natural and Social Sciences. Proceedings of the Boston Colloquium for the Philosophy of Science 1969–1972.* 1974.
15. Robert S. Cohen, J. J. Stachel, and Marx W. Wartofsky (eds.), *For Dirk Struik. Scientific, Historical and Political Essays in Honor of Dirk Struik.* 1974.
16. Norman Geschwind, *Selected Papers on Language and the Brain.* 1974.
17. B. G. Kuznetsov, *Reason and Being: Studies in Classical Rationalism and Non-Classical Science.* (forthcoming).
18. Peter Mittelstaedt, *Philosophical Problems of Modern Physics.* 1976.
19. Henry Mehlberg, *Time, Causality, and the Quantum Theory* (2 vols.). 1980.
20. Kenneth F. Schaffner and Robert S. Cohen (eds.), *Proceedings of the 1972 Biennial Meeting, Philosophy of Science Association.* 1974.
21. R. S. Cohen and J. J. Stachel (eds.), *Selected Papers of Léon Rosenfeld.* 1978.

22. Milic Capek (ed.), *The Concepts of Space and Time. Their Structure and Their Development.* 1976.
23. Marjorie Grene, *The Understanding of Nature. Essays in the Philosophy of Biology.* 1974.
24. Don Ihde, *Technics and Praxis. A Philosophy of Technology.* 1978.
25. Jaakko Hintikka and Unto Remes. *The Method of Analysis. Its Geometrical Origin and Its General Significance.* 1974.
26. John Emery Murdoch and Edith Dudley Sylla, *The Cultural Context of Medieval Learning.* 1975.
27. Marjorie Grene and Everett Mendelsohn (eds.), *Topics in the Philosophy of Biology.* 1976.
28. Joseph Agassi, *Science in Flux.* 1975.
29. Jerzy J. Wiatr (ed.), *Polish Essays in the Methodology of the Social Sciences.* 1979.
30. Peter Janich, *Protophysics of Time.* 1985.
31. Robert S. Cohen and Marx W. Wartofsky (eds.), *Language, Logic, and Method.* 1983.
32. R. S. Cohen, C. A. Hooker, A. C. Michalos, and J. W. van Evra (eds.), *PSA 1974: Proceedings of the 1974 Biennial Meeting of the Philosophy of Science Association.* 1976.
33. Gerald Holton and William Blanpied (eds.), *Science and Its Public: The Changing Relationship.* 1976.
34. Mirko D. Grmek (ed.), *On Scientific Discovery.* 1980.
35. Stefan Amsterdamski, *Between Experience and Metaphysics. Philosophical Problems of the Evolution of Science.* 1975.
36. Mihailo Marković and Gajo Petrović, *Praxis. Yugoslav Essays in the Philosophy and Methodology of the Social Sciences.* 1979.
37. Hermann von Helmholtz, *Epistemological Writings. The Paul Hertz/Moritz Schlick Centenary Edition of 1921 with Notes and Commentary by the Editors.* (Newly translated by Malcolm F. Lowe. Edited, with an Introduction and Bibliography, by Robert S. Cohen and Yehuda Elkana.) 1977.
38. R. M. Martin, *Pragmatics, Truth, and Language.* 1979.
39. R. S. Cohen, P. K. Feyerabend, and M. W. Wartofsky (eds.), *Essays in Memory of Imre Lakatos.* 1976.
42. Humberto R. Maturana and Francisco J. Varela, *Autopoiesis and Cognition. The Realization of the Living.* 1980.
43. A. Kasher (ed.), *Language in Focus: Foundations, Methods and Systems. Essays Dedicated to Yehoshua Bar-Hillel.* 1976.
44. Trân Duc Thao, *Investigations into the Origin of Language and Consciousness.* (Translated by Daniel J. Herman and Robert L. Armstrong; edited by Carolyn R. Fawcett and Robert S. Cohen.) 1984.
46. Peter L. Kapitza, *Experiment, Theory, Practice.* 1980.
47. Maria L. Dalla Chiara (ed.), *Italian Studies in the Philosophy of Science.* 1980.
48. Marx W. Wartofsky, *Models: Representation and the Scientific Understanding.* 1979.
49. Trân Duc Thao, *Phenomenology and Dialectical Materialism.* 1985.
50. Yehuda Fried and Joseph Agassi, *Paranoia: A Study in Diagnosis.* 1976.
51. Kurt H. Wolff, *Surrender and Catch: Experience and Inquiry Today.* 1976.
52. Karel Kosík, *Dialectics of the Concrete.* 1976.
53. Nelson Goodman, *The Structure of Appearance.* (Third edition.) 1977.

54. Herbert A. Simon, *Models of Discovery and Other Topics in the Methods of Science*. 1977.
55. Morris Lazerowitz, *The Language of Philosophy. Freud and Wittgenstein*. 1977.
56. Thomas Nickles (ed.), *Scientific Discovery, Logic, and Rationality*. 1980.
57. Joseph Margolis, *Persons and Minds. The Prospects of Nonreductive Materialism*. 1977.
59. Gerard Radnitzky and Gunnar Andersson (eds.), *The Structure and Development of Science*. 1979.
60. Thomas Nickles (ed.), *Scientific Discovery: Case Studies*. 1980.
61. Maurice A. Finocchiaro, *Galileo and the Art of Reasoning*. 1980.
62. William A. Wallace, *Prelude to Galileo*. 1981.
63. Friedrich Rapp, *Analytical Philosophy of Technology*. 1981.
64. Robert S. Cohen and Marx W. Wartofsky (eds.), *Hegel and the Sciences*. 1984.
65. Joseph Agassi, *Science and Society*. 1981.
66. Ladislav Tondl, *Problems of Semantics*. 1981.
67. Joseph Agassi and Robert S. Cohen (eds.), *Scientific Philosophy Today*. 1982.
68. Władysław Krajewski (ed.), *Polish Essays in the Philosophy of the Natural Sciences*. 1982.
69. James H. Fetzer, *Scientific Knowledge*. 1981.
70. Stephen Grossberg, *Studies of Mind and Brain*. 1982.
71. Robert S. Cohen and Marx W. Wartofsky (eds.), *Epistemology, Methodology, and the Social Sciences*. 1983.
72. Karel Berka, *Measurement*. 1983.
73. G. L. Pandit, *The Structure and Growth of Scientific Knowledge*. 1983.
74. A. A. Zinov'ev, *Logical Physics*. 1983.
75. Gilles-Gaston Granger, *Formal Thought and the Sciences of Man*. 1983.
76. R. S. Cohen and L. Laudan (eds.), *Physics, Philosophy and Psychoanalysis*. 1983.
77. G. Böhme et al., *Finalization in Science*, ed. by W. Schäfer. 1983.
78. D. Shapere, *Reason and the Search for Knowledge*. 1983.
79. G. Andersson, *Rationality in Science and Politics*. 1984.
80. P. T. Durbin and F. Rapp, *Philosophy and Technology*. 1984.
81. M. Marković, *Dialectical Theory of Meaning*. 1984.
82. R. S. Cohen and M. W. Wartofsky, *Physical Sciences and History of Physics*. 1984.
83. E. Meyerson, *The Relativistic Deduction*. 1985.
84. R. S. Cohen and M. W. Wartofsky, *Methodology, Metaphysics and the History of Sciences*. 1984.
85. György Tamás, *The Logic of Categories*. 1985.
86. Sergio L. de C. Fernandes, *Foundations of Objective Knowledge*. 1985.
87. Robert S. Cohen and Thomas Schnelle (eds.), *Cognition and Fact*. 1985.
88. Gideon Freudenthal, *Atom and Individual in the Age of Newton*. 1985.
89. A. Donagan, A. N. Perovich, Jr., and M. V. Wedin (eds.), *Human Nature and Natural Knowledge*. 1985.
90. C. Mitcham and A. Huning (eds.), *Philosophy and Technology II*. 1986.
91. M. Grene and D. Nails (eds.), *Spinoza and the Sciences*. 1986.
92. S. P. Turner, *The Search for a Methodology of Social Science*. 1986.
93. I. C. Jarvie, *Thinking About Society: Theory and Practice*. 1986.
94. Edna Ullmann-Margalit (ed.), *The Kaleidoscope of Science*. 1986.
95. Edna Ullmann-Margalit (ed.), *The Prism of Science*. 1986.
96. G. Markus, *Language and Production*. 1986.